Babak Ebrahimi

Electromagnetic Dampers in Vehicle Suspension Systems

Babak Ebrahimi

Electromagnetic Dampers in Vehicle Suspension Systems

Design, modeling, and implementation

VDM Verlag Dr. Müller

Impressum/Imprint (nur für Deutschland/ only for Germany)
Bibliografische Information der Deutschen Nationalbibliothek: Die Deutsche Nationalbibliothek verzeichnet diese Publikation in der Deutschen Nationalbibliografie; detaillierte bibliografische Daten sind im Internet über http://dnb.d-nb.de abrufbar.

Coverbild: www.purestockx.com

Verlag: VDM Verlag Dr. Müller Aktiengesellschaft & Co. KG
Dudweiler Landstr. 99, 66123 Saarbrücken, Deutschland
Telefon +49 681 9100-698, Telefax +49 681 9100-988, Email: info@vdm-verlag.de
Zugl.: Waterloo, University of Waterloo, 2009

Herstellung in Deutschland:
Schaltungsdienst Lange o.H.G., Berlin
Books on Demand GmbH, Norderstedt
Reha GmbH, Saarbrücken
Amazon Distribution GmbH, Leipzig
ISBN: 978-3-639-21394-2

Imprint (only for USA, GB)
Bibliographic information published by the Deutsche Nationalbibliothek: The Deutsche Nationalbibliothek lists this publication in the Deutsche Nationalbibliografie; detailed bibliographic data are available in the Internet at http://dnb.d-nb.de .

Cover image: www.purestockx.com

Publisher:
VDM Verlag Dr. Müller Aktiengesellschaft & Co. KG
Dudweiler Landstr. 99, 66123 Saarbrücken, Germany
Phone +49 681 9100-698, Fax +49 681 9100-988, Email: info@vdm-publishing.com

Printed in the U.S.A.
Printed in the U.K. by (see last page)
ISBN: 978-3-639-21394-2

o

o o 'o

O

Dvdicaflon(

ObKCZAAKACS AAK'ASCA[KCR[zACEbKAA'm'zCmDCK[AKCEAK[SszCO Abs[CSS[AKfAsSSz EEbKSCA[KCa[cb[K'Sb[AlO
lbv[Cffz[ll[KCmDCKK'v[CSbCSA[Cl[KO

O

O O ”O

Tablv(of(ConRvnB(

O

O

O

O

O

LisRof(Lig, rvs(

O

O

O

O O

O

L

L

L

LisRof(Tablvs(

Nomenclature

e_g	laTeral area of The ole
e_g	-roee ee-Tion area of The ma-neT
e_g	-roee ee-Tion area of The ieTon
e_g	-roee ee-Tion area of The rod
	ma-neTi- flul deneiTy
e_g	air--a flul deneiTy
e_g	radial flul deneiTy
e_g	al ial flul deneiTy
e_{ggg}	remanenTflul deneiTy
e	dam in- -oeffi-ienT
e_g	dam in- -oeffi-ienTin -om reeeion
e_g	ai era-e dam in- -oeffi-ienT
e_g	dam in- -oeffi-ienTin el Teneion
$e_{gg} e\ e_{gg}\ ee_{gg}\ ee_{gg}$	die-har-e -oeffi-ienT
$e_{g\,ax}$	mal imum dam in- for-e
e_g	aeeii e dam in- -oeffi-ienT
e_{gkx}	ideal eOyhooOdam in- -oeffi-ienT
e_g	hydrauli- diameTer
e	rod diameTer
e_g	de Th
e	ToTal indu-ed ele-TromoTi e for-e
e_{kgaag}	Taneformer ele-TromoTi e for-e
$e_{gkkkaal}$	moTional ele-TromoTi e for-e
$e\,tt\,a$	elli Ti- inTe-ral of The fireTOnd
ee	for-e
e_a	a-Ti e for-e
e_g	-riTi-al bu-Olin- for-e
e_g	dam in- for-e
e_g	fri-Tion for-e
e_g	ToTal hybrid dam in- for-e
e_g	eTaTi- for-e
e	fri-Tion fa-Tor
$e\ e$	L raehof number
e	-a beT een T o ma-neTe
e_g	-oer-ii e ma-neTi- field

e_{gxk}	el Ternal ma- neTaaTon
e_g	for- ed--oni e-Ton heaTTanefer -oeffi-ienT
e_g	naTural--oni e-Ton heaTTanefer -oeffi-ienT
e	ele-Ti-al -urrenT
e	eaO-urrenT
e_g	momenTof inerTa
e	-urrenTdeneiTy
g	equii alenTi olume -urrenTdeneiTy
g	equii alenTeurfa-e -urrenTdeneiTy
t_g	-urrenTdeneiTy aTThe eurfa-e
e	Thermal -ondu-Ti iTy
e_g	ThrueT-oneTanT
e_{gg}	ele-Ti- loadin-
$e\, tt\, a$	elli Ti- inTe-ral of The ee-ond Oind
t	ToTal eue eneion eTiffneee
t_a	airle Thermal -ondu-Ti iTy
t_g	-ondu-Torle Thermal -ondu-Ti iTy
t_g	moi erle Thermal -ondu-Ti iTy
t_g	eue eneion eTiffneee
t_g	Tre eTiffneee
e	indu-Tan-e
e_k	free bu-Olin- len-Th
l_g	ma- neTe len-Th
e	ma- neTaaTon
g	remanenTma- neTaaTon
e	maee
e_g	e run- maee
e_g	une run- maee
e	number of Turne in a -oil
e_g	number of -oile
$e\, e$	TueeelTnumber
a	fa-Tor of eafeTy
$\hat{n}$	uniTeurfa-e normal
e	ol er loee
e_g	-om reeeion -hamber reeeure
e_g	el Teneion -hamber reeeure
e_g	hydrauli- reeeure
e_g	fooT-hamber reeeure

e_g	LrandTI number
e	number of olee
$e_{gg} e\ e_{gg}\ ee_{gg}\ ee_{gg}$	i olumeTri- flol raTE
e	reeieTan- e
ee	Teynolde number
e_g	ermanenTma- neTradiue
e_{gag}	equii alenTrelu- Tan- e of The air-- a
e_{gag}	equii alenTrelu- Tan- e of The ma- neT
e	elemenTlen- Th
e_{aa}	-ondu- Tor ineide radiue
e_{kgk}	-ondu- Tor ouTeide radiue
e	Teneile eTen- Th
e_{gg}	eTaTor -oille -roee ee- Tion area
e_g	enduran- e limiT
e	rod radiue
e_g	air-- a Thi-Cneee
e	eriod
e_g	-ondu- Torle eurfa- e Tem eraTure
e_g	ma- neTTem eraTure
t	Time
	i elo- iTy
e_g	el Teneion i elo- iTy
e_g	-om reeeion i elo- iTy
e_g	-oil l idTh
g	relaTi e die la- emenT
g	e run- maee die la- emenT
M_w	e run- maee i elo- iTy
M_w	e run- maee a- -eleraTion
e	baee die la- emenTam liTude
	baee die la- emenT
M	baee i elo- iTy
R	lon- iTudinal uniTi e- Tor
e	elemenTlen- Th
e_g	dieTan- e beTI een TI o -urrenT -arryin- loo e
β	i olume el aneion -oeffi- ienT
β	-ondu- Tor i olume
β	die la- emenTTanemieeibiliTy
βel_g	-ondu- Tor Thi-Cneee

β_g	de Th of eneTaTon
β_g	eTaTor - ore Thi-Qheee
β_g	-ylinder Thi-Qheee
β	dam in- raTo
β	a--eleraTon Tanemieeibility
β	an-le and haee
β_g	l indin- Tem eraTure riee
β	relaTi e die la-emenTTanemieeibiliTy
β_g	l indin- fa-Tor
β_{gx}	flul linQa-e due To a ma-neT
β	fluid i ie-oeiTy
β_a	abeoluTe ermeabiliTy
β_k	ermeabiliTy of free e a-e
β_g	relaTi e ermeabiliTy
β_{ggg}	re-oil ermeabiliTy
β	fluid deneiTy
β	-ondu-Ti iTy
β_g	hoo eTeee
β	ole iTh
β_g	ele-Tri-al Tme -oneTanT
β_g	ma-neTThi-Qheee
β_g	-oil iTh
υ	QnemaTi- i ie-oeiTy
β	haee an-le
ϕR	Tan-enTal uniTi e-Tor
υ_Y	ma-neTi- flul in The -a
υ_Y	ma-neTi- flul due To The ma-neTe
β	frequen-y
β_a	naTural frequen-y

Ch1pT bh1h

InTbPdC4TBPnh

Lue eneion eyeTemeTin The auTbmoTi e a li-aTon -onTel TThai e been deei-ned Tb mainTain -onTa-T beT een a i ehi-lele Tree and The roadTand Tb ieolaTe The frame of The i ehi-le from road dieTurban-eem n am ereTor eo--alled eho-O abeorbereTae The undeniable hearTof eue eneion eyeTemeTredu-e The effe-T of a eudden bum by emooThin- ouT The eho-OmIn moeT eho-O abeorbereT The ener-y ie -oni erTed inTb heaTi ia i ie-oue fluidmIn hydrauli- -ylindereTThe hydrauli- fluid ie heaTed u mIn air -ylindereTThe hoTair ie emiTTed inTb The aTmoe heremThere are eei eral -ommon a roa-hee for eho-O abeor TionTin-ludin- maTerial hyeTereeieTdry fri-TionTfluid fri-TionT-om reeeion of -aeTand eddy -urrenTem

In The aeTfel yeareTim roi emenTe in ol er ele-Troni-e and ma-neT- maTeriale hai e reeulTed in ei-nifi-anTim roi emenTe in ele-To-me-hani-al dei i-eeTredu-in- Their i olumeTl ei-hTTand -oeTTand im roi in- Their oi erall effi-ien-y and reliabiliTymTheee dei elo menTe jueTify an analyeie and im lemenTaTion of ele-Toma-neT- dei i-ee in eue eneion eyeTememIn moeTdam ereTThe ener-y ie -oni erTed inTb heaTand dieei aTed l iThouTbein- ueed; in ele-Toma-neT- dam ereTThe dieei aTed ener-y -an be eTbred ae ele-Tri-al ener-y and ueed laTermThe uee of ele-Toma-neT- dam ere in eue eneion eyeTeme hae eei eral benefiTe -om ared Tb hydrauli-T neumaT-Tor oTher me-hani-al dam eremLle-Toma-neT- dam ere -an fun-Tion eimulTaneouely ae eeneore and a-TuaTbremThe e rin-effe-T -an be added Tb The eyeTem by meane of ele-Toma-neTeT ol ered by LermanenT Ma-neTe (LMe2mMoreoi erT ele-Toma-neT- dam ere -an l orO under i ery lol eTaT- fri-TionmLereT The dam in- -oeffi-ienTie -onTrolled ra idly and reliably Throu-h ele-Tri-al mani ulaTionem

i lThou-h a-Ti e eue eneione hai e eu erb erforman-eTThey hai e hi-h ener-y -oneum TionTare noT fail-eafe in -aee of ol er failureTand are heai y and el eneii e -om ared Tb -oni enTional i ariable dam ereTuein- oTher Te-hnolo-iee eu-h ae ele-Toma-neT--i ali e and Ma-neTb-Theolo-i-al (MT2 fluid dam eremLybrid ele-Toma-neT- dam ereTl hi-h are ro oeed in Thie manue-ri TTare oTenTal eoluTione Tb The hi-h l ei-hTThi-h -oeTTand fail-eafeTy ieeuee in an a-Ti e eue eneion eyeTemmThe hybrid ele-Toma-neT- dam er -ombinee The erforman-e and -onTrollabiliTy of an a-Ti e ele-Toma-neT- dam er l iTh The reliabiliTy of aeeii e dam ere in a ein-le a-Oa-eTeai in- l ei-hTand -oeTT and maOn- The dam er fail-eafe in -aee of ol er failuremLin-e The a-Ti e ele-Toma-neT- dam er uTliaee ma-neT- -om onenTETone oTEnTal deei-n for eu lyin- The aeeii e dam in- effe-T ie The eddy -urrenTdam in- henomenonm

Lddy -urrenTE are indu-ed in a -ondu-Tor eiTher by The moi emenTof The -ondu-Tor in a eTaT- field or a -han-e in The eTen-Th of The ma-neT- fieldmThe -eneraTEd eddy -urrenTE -reaTE ele-Toma-neTE l iTh ma-neT- fielde ThaTo oee The -han-e in an el Ternal ma-neT- fieldT-auein- a re uleii e for-e ro orTional To The relaTi e i elo-iTy of The field and -ondu-Tormi oneequenTlyT The eddy -urrenT dam in- dei i-e a-TE liOe a i ie-oue dam erT-auein- The i ibraTion ener-y of The moi in- maee To dieei aTE Throu-h reeieTi e/Joule heaTin-T-eneraTEd in The -ondu-Tin- -om onenTmThe uee of The eddy -urrenT dam in- henomenon ie a ealin-Tbe-auee The dam er eTu-Ture ie eim le and requiree neiTher el Ternal ol er eu ly nor ele-Toni- dei i-ee; in addiTionTno fluid ie ini oli ed in The dam erT and The moi in- arTE of The dam er hai e no me-hani-al -onTa-TmThe eddy -urrenT dam in- henomenon ie die-ueeed in more deTail in i ha Ter 3m

Thie InTodu-Tion -ha Ter ouTinee Three main eubje-TmaTTerenFireTTThe i ehi-le eue eneion eyeTEme are inTodu-edTfo-uein- on The differenTeue eneion eyeTEm Ty ee and -oni enTional -onTol eTaTE-iee ini oli ed in Theee eyeTEmemLe-ondT differenT i ehi-le eue eneion modele ae l ell ae The i ehi-le eue eneion erforman-e indel ee are die-ueeedmFinallyTdam ere ae The main fo-ue in Thie TEl Tare -oneideredTThe -oni enTional and The noi el ro oeed TE-hnolo-iee are inTodu-edTand The liTEraTure ie rei iel edmThe InTodu-Tion -ha Ter ie finaliaed l iTh a brief eummary of moTi aTionTobje-Ti eeTand The booOouTinem

3.3 V. h64-. IbCsp. 5s665IbysT 0 h

The rimary fun-Tion of The i ehi-le eue eneion eyeTEm ie To roi ide a -omforTable rideTThrou-h ieolaTion of The i ehi-le body from road irre-ulariTeeTand enhan-e The ride handlin- by rodu-in- a -onTinuoue road-l heel -onTa-Tm

i --ordin- To The lei el of -onTollabiliTyTeue eneion eyeTEme are -laeeified ae aeeii eTa-Ti eTor eemi-a-Ti emi eide from The inherenTadi anTa-ee and dieadi anTa-ee of ea-h Ty eTall of Them uTliae The e rin- and dam er uniTEmTie i erified ThaTThe aeeii e and eemi-a-Ti e eue eneion eyeTEme -annoT roi ide The aboi e-menTioned requiremenTE eaTefa-TorilyTbe-auee of The -onTaeTin- naTure of Theee requiremenTemi -Tii e eue eneion in iTe i arioue forme offere a meane Tb relal Thoee -om romieee (a endel *et al.*T5((52mIn a-Ti e eue eneion eyeTemeTa-TuaTbreTl hi-h are o eraTin- by el Ternal ol erT are ueed Tb -eneraTe The ne-eeeary for-e Tb eu reee The i ibraTionemL ol ei erTa-Ti e eue eneione hai e dral ba-Œ re-ardin- Their eiaeTl ei-hTT ol er -oneum TonTand fail-eafeTy ieeueem

Lemi-a-Ti e eue eneione aleo -onTain e rin- and dam in- elemenTemThe ro erTee of Theee elemenTe -an be -onTolled el Ternally by eu lyin- a ei-nal or oTher Ty ee of el Ternal ol er Tb The eyeTemmIn addiTionTheee -hara-TerieTi-e -an be -han-ed eo ThaTdifferenTlei ele of The reeieTn- for-ee are rodu-edT a--ordin- Tb a lol - ol er -onTol ei-nalmTheee eyeTeme arTally benefiT from The adi anTa-ee of a-Ti e eyeTeme l iTh The reliabiliTy of aeeii e eyeTememIf The -onTol eyeTem faileTThe eemi-a-Ti e eyeTem -onTinuee Tb o eraTe in The aeeii e modem

There are eei eral qualiTaTi e meaeuree ThaTmueTbe -oneidered by The eue eneion eyeTem deei-ner in order Tb a-hiei e The deeired erforman-eTin-ludin-3ride -omforT(-orree onde Tb The a--eleraTion felTby The aeeen-er2Troad-handlin-TThe relaTi e moTion beTT een The i ehi-le body and l heeleTand The body moTione durin- -riTi-al maneui ere eu-h ae braOin- and -ornerin- (2 i-larbe-ian *et al.*T211 82m The i ehi-le eue eneion erforman-e indel ee are die-ueeed in Le-Tion 5m2m2 in -reaTer deTailm

3.3.3 63ss6v. lbCsp. 5s655lbysT0 sh

Laeeii e eue eneion eyeTeme are The moeT-ommon eyeTeme ThaTare ueed in -ommer-ial aeeen-er -aremThey are -om oeed of -oni enTional e rin-eTand ein-le or TT in-Tube oil dam ere l iTh -oneTanT dam in- ro erTeem The deei-ner re-eeTe The fil dam in- ro erTee Tb a-hiei e o Timum erforman-e for The inTended a li-aTionmThe dieadi anTa-e of aeeii e eue eneion eyeTeme l iTh -oneTanTdam in- -hara-TerieTi-e ie ThaTeeTTin- The deei-n arameTere ie a -om romiee beTT een The ride qualiTy and handlin-mLenerallyTeofTer dam ere roi ide a more -omforTable rideTl hile eTiffer onee roi ide beTTer eTabiliTy and Thue beTTer road-handlin-mLaeeii e eue eneion eyeTeme are Tuned a--ordin- Tb The el e-Ted o eraTin- -ondiTioneTbuTa -om romiee ie all aye made beTT een deei-nin- for The T o o oein- -oalemThereforeTThe erforman-e in ea-h area ie limiTed a--ordin- Tb Thie -om romiee (Lillee ieT211 L2mL ol ei erTTradiTional aeeii e eue eneion eyeTeme are lol -oeTTand relaTi ely eim le Tb manufa-Turem

3.3.0 b. 0 6-34Tv. lbCsp. 5s665lbysT 0 sh

Lin-e fireT ro oeed by i roeby and ; arno (5(f32Teemi-a-Ti e eue eneion eyeTme -onTnue To -ain o ulariTy in i ehi-le eue eneion eyeTm a li-aToneTdue To Their adi anTa-eoue -hara-TerieTi-e oi er aeeii e eue eneion eyeTmemLemi-a-Ti e (aleo Qnol n ae ada Ti e- aeeii e2 eue eneion eyeTme are eeeenTally aeeii e eyeTme in l hi-h The dam in- ro erTee -an be adjueTed To eome el TenTmThueT eemi-a-Ti e eue eneion eyeTme el Tend The oeeible ran-e of dam in- -hara-TerieTi-e obTainable from a aeeii e dam er (i arTerT5((82T(2 araOT5((22Tand (; ooT*et al.*T21 l 42mThe dam in- -hara-TerieTi-e of a eemi-a-Ti e dam er -an be adjueTed Throu-h a lyin- a lol - ol er ei-nalmLemi-a-Ti e eyeTme are a -om romiee beT een The a-Ti e and aeeii e eyeTme (i lanoly *et al.*T 5(8L2mThey are -ommer-ialiaed re-enTy by meane of eiTher a eolenoid i ali e ae an adjueTable orifi-eTor MT-fluid dam ereTboTh of l hi-h are i ery el eneii emLeadin- auTomoTi e manufa-Turere eu-h ae LM and n oli o hai e eTarTed The im lemenTaTon of Theee eemi-a-Ti e eue eneion eyeTme for Their hi-h-end auTomobileemL ol ei erTThere el ieTmany -hallen-ee ThaT mueTbe oi er-ome for Theee Te-hnolo-ieem The MT dam ere hai e eTll eome -ru-ial ieeueeTeu-h ae MT de-radaTon l iTh TmeT Tem eraTure eeneiTi iTyTand eealin- roblemem

3.3.3 A4Tv. lbCsp. 5s665lbysT 0 sh

i -Ti e eue eneion eyeTm refere To a eyeTm ThaT ueee an a-Ti e ol er eour-e To a-TuaTe The eue eneion linQe by el Tendin- or -onTa-Tn- Them ae required (L illee ieT21 l L2mThe main -hallen-e in The auTomobile eue eneion eyeTm deei-n ie The Trade-off beT een The -onfli-Tn- requiremenTe of ride--omforTand handlin-nIn The aeTfel de-adeeTdue To adi an-ee in eeneore/a-TuaTore Te-hnolo-ieeT a-Ti e eue eneion eyeTme hai e emer-ed ae an a-Ti e reeear-h field To addreee Thoee -om romieeem

In an a-Ti e eue eneionT-onTrolled for-ee are inTrodu-ed To The eue eneion by meane of hydrauli- or ele-Tri- a-TuaToreTbeT een The e run- and une run--maee of The l heel aeeembliee (2 rol nT21 l 52m ; arno (5(832 eTudied The effe-T of addin- an a-Ti e dam in- for-e To The eue eneion eyeTmm MilliQen (5(882 dei elo ed The fireT fully a-Ti e eue eneion To be ueed in a eue eneione eyeTm a li-aTonmi i ariable for-e ie roi ided by The a-Ti e eue eneion aT ea-h l heel To -onTnuouely modify The ride and handlin- -hara-TerieTi-emThe Qey differen-e beT een eemi-a-Ti e and a-Ti e eue eneion eyeTme ie ThaT The laTTer a liee an el Ternal for-e To The i ehi-le body eiTher in an u l ard or dol nl ard dire-TonT re-ardleee of The abeoluTe i ehi-le body i elo-iTymi -Ti e eue eneion oi er-omee The -om romieee required in Tunin- aeeii e dam eremLeneoreTae eeeenTal elemenTe of an a-Ti e eue eneionTare ueed To meaeure The eue eneion moi emenTe aTdifferenT oinTem lThou-h a-Ti e eue eneione hai e eu erb erforman-eTThe ra-Ti-al im lemenTaTion of a-Ti e eue eneion in i ehi-lee hae been limiTed due To Their hi-h l ei-hTT-oeTT ol er -oneum TionTand redu-ed reliabiliTy ThaTare noTjueTified by The limiTed in-remenTal benefiTTo The aeeen-er (n eoT21 l f2mMoeTof The eue eneion eyeTeme ThaTare ro oeed and im lemenTed in i ehi-le eue eneion eyeTeme are baeed on hydrauli- or neumaTi- dei i-eem

3.0 b(0 p-ff6 dlV. h64-. lV 6d.-sl35dl6. bf6b0 354. lV5d. x. sh

ITie oeeible To eTudy The fundamenTal -on-e Te of i ehi-le dynami-e baeed on eim lified i ehi-le modelemThe -hoi-e of The eue eneion eyeTem modele de ende on The i ibraTion mode and The eim li-iTy of The analyeie ini oli edm2 aei- i ehi-le modele ThaThai e been rei iel ed in The liTeraTure are The quarTer--ar one-de-ree-of-freedom (5n OF2 model To eTudy The boun-e modeT The quarTer--ar 2n OF model To eTudy The l heel-ho ae l ell ae The boun-e modeT The half--ar 4n OF model for boun-e and iT-h modeeTand The full--ar fn OF model for boun-eT iT-h and roll modee (n eom21 l f2m The moeT-eneral and -oneTru-Ti e auTomoTi e eue eneion eyeTem deei-n informaTion derii ee from a ein-le-l heel eTaTionT quarTer--ar re reeenTaTion (Lhar and i rollaT 5(8f2mQuarTer--ar modele are erfe-Tfor a i ehi-le l iTh a lon-iTudinally inde endenTeue eneion and a eymmeTri- maee dieTribuTion in The i ehi-le bodyTl here The fronTand rear eue eneion ie de-ou led (MiTe-hOeT5(L52mThie model ie a -ood a rol imaTion of a real eue eneion eyeTem only for lon- l ai elen-Th and lol frequen-y in uTe (Lhar and LaeeanT5(842mITdoee noTre reeenTThe -eomeTri- effe-Te of hai in- four l heeleTand doee noTin-or oraTe The lon-iTudinal inTer-onne-TionmL ol ei erTiT-onTaine The moeTfundamenTal feaTuree of The real eue eneion eyeTemTeu-h ae eue eneion eyeTem for-eeTl hi-h are ro erly a lied beTl een a i ehi-lele l heele and bodymITie The eim leeTmodel l hi-h hae Theee feaTureeTreTainin- adi anTa-ee oi er more -om lel modele in Terme of hai in- fel er erforman-e arameTeremThere ie only a ein-le in uTini oli ed in a quarTer--ar model ThaTeim lifiee The erforman-e -om uTaTion and The a li-aTion of o Timal -onTrol eTaTe-ieemIn addiTionTThe quarTer--ar model offere a -om reheneii e model for undereTandin- The -orrelaTion beTl een erforman-e and deei-n arameTeremThe quarTer--ar model l iTh realieTi- road in uTe e-Ta ie noTeeneiTi e To The body-To-l heel maeeTand Tire eTiffneee-To-l ei-hT eu orTed raTioe (TybaT5(f42 and (Lhar and LaeeanT5(842T roi ided ThaT Theee are reaeonably eele-Tedm

2ernard *et al.* (5((52 ro oeed a roll-oi er model To redi-T The i ehi-le roll -hara-TerieTi-em Lol ei erTiTe a li-aTon ie limiTed be-auee The model -annoT redi-TThe i erTi-al moTon of The i ehi-le (LelaminaeabT21 1 82mIn order To -oneider The rolloi er ae l ell ae The i erTi-al moTioneTmore -om lel modele are ro oeedT eu-h ae a 4n OF half--ar i erTi-al dynami- model ThaT hae already been dei elo ed and ueed in The numeroue reeear-h l orOe (L roi aTT5((f 2 and (; yon- eu *et al.* T21 1 f 2m

3.0.3 b 5. lB5dhTw6IDb VIQC3bT b-C3bIV 6d. -sh

The dynami- behai ior of i ehi-le i erTi-al ieolaTion ro erTiee -an be re reeenTed and analyaed by one or T o n OF quarTer--ar modeleTae ehol n in Fi- m5-5 (LelaminaeabT21 1 82m

Vfj)l Vk FpquWRrggWp ods'p bBBVk FpquWRrggWp ods'p

V(0)87-7HO58I75dHIw6IDOVIqC7OBO47(0)0 6d8-s8h

In Thie model re reeenTaTionT e_g ie a quarTer of The l hole i ehi-le body maee (*eeeeae-e aee*2Tl hile The l heel maee ie re reeenTed by e_g and -alled The *eaeeeeae-e aeem* t_g and t_g re reeenTThe eue eneion and Tire eTiffneeeTree e-Ti elynThe -oi ernin- equaTion of moTon for The 5n OF eyeTem in Fi- m5-5(a2 ie

$$+_{+}\mathcal{M}\,\sigma\,\sigma J9\mathcal{M}\,\sigma\,\mathcal{M}B\sigma+_{+}9\mathcal{M}\,\sigma\,\mathcal{A}B\sigma+_{)}\,T \qquad (1\text{-}52$$

l here t ie The ToTal eue eneion eTiffneeeTand e_a ie The a-TuaTor a-Ti e for-e in -aee of The a-Ti e eue eneion eyeTemTand ie eeT by The -onTrol eTraTe-yme_{gx}ie The aeeii e dam in- -oeffi-ienT in The eyeTemnIn addiTonT $_g$ and are The maee moi emenTe and The road dieTurban-eeri oneiderin- The road dieTurban-e ae a einueoidal fun-Tion ($\mathcal{M}\sigma + +))\,9\upsilon J B2$T The equaTion of moTon (1 -52 ie el reeeed ae (LelaminaeabT21 1 82

$$\mathcal{M}\zeta\,B\zeta\zeta_{)}\,\mathcal{M}\zeta\,\zeta_{)}{}^{B}\mathcal{M}\sigma+\zeta^{B}sA9\zeta JB\sigma+_{)}\,/\,+_{+}\mathrm{Rp} \qquad 9'\,gBBp$$

p

' hsrspRhsppWW sRrspWspdW pAVgprVRbop9ζ BpVVRurWp7rsqusVgyp9ζ) BpWdpRhsprs'WRAAspdAp'Vgs. sVRp 9M_wBpWspds7AVsdpWp

p $$\zeta \, \sigma \frac{+}{B\sqrt{J+_w}} R\zeta_) \, \sigma \sqrt{\frac{J}{+_w}} R \; M_w \sigma M_w \sigma M_f \qquad 9' g Bp$$

k VpRhspoRhsrphWdRpRhspgoAsrVAVgpsquVRboVp7orpWpBVk FpsuspsVsAbVpsysR. pAVpFAg.plgl9bBpAsp

p $$\begin{aligned} &+_+ M_f \sigma \sigma J_+ 9M_f \sigma M_f B \sigma +_+ 9M_f \sigma M_f B \sigma +_) R \\ &+_+ M_f \sigma J_+ 9M_f \sigma M_f B \sigma J_+ 9M_f \sigma M_f B \zeta +_+ 9M_f \sigma M_f B \zeta +_) . \end{aligned} \quad p \qquad 9' g4Bp$$

AspWgo. . oVpprVgRgsRpRhspl Vk FpquWRrggWp ods'phWpbssVpAWRypussdpWpWpo' sr7u'pWdpsA p'sp . sRhodpRopWdrssspRhspAsrRgWp oRbVpo7pAshAg'ssRpsughpWpAVp9; WVopppt*JMl.*Rpl 9E9BRp9; WVopppt*JMl.*Rp 1974BRp9RWhsjVpt*JMl.*Rpl 981 BRp9A'Wo'ypt*JMl.*Rpl 98EBRp9EhsVRpB' ' 1BpWdp9Es'W AVWWbRpB' ' 8BpAspoVsp o7pRhspsubjsgRspAVpRhAspsRudypAspRopWdopRpVpsA p'sp. ods'pRopWWyWpRhspdW pAVgpghWVgRrAsRgsp gorrs'WRbVp' AbpRhspsuspsVsAbVppsr7or. WgsRpWpl Vk FpquWRrggWp ods'pAsprA WAypWdopRdpAVpRhAsp book.pp

3.0.0 V. h64-. lbCsp. 5s665l6. bf6b0 354. lN5d. x. sh

RhsrspWspWVu. bsrpo7pAVds' sspAVpRhspsuspsVsAbVpgoVRo'pWWysAspWdrsssAVgpRhsppsr7or. WgspRo7pRhsp suspsVsAbVpsysR. .pRAdspgo. 7orRpWdproWdghWd'AVgpWspW oVgpRhsp osRpgo. . oV'ypussdpAVds' sspAVpWp AshAg'spsuspsVsAbVpgoVR' RpMughprsssWghphWpbssVpgoVdugRdpRop7AVdp7VgRorspRhWRpW7sgRprAdspgo. 7orRp 9Eo''WdpWdpEA oVsRpl 984BRp9Vs. Agpt*JMl.*RpB' ' BBpWdp9lhsWpt*JMl.*RpB' ' 8BpRhsp7or. srRp7orps' W p'sRp AVAssRgWRdphu. WpbshWAorpuVdsrprWdo. pAAorWRbVsRpWdpds. oVsRWRspRhWphu. WspWspVor. W'yp AsrypssVsAAAspRop7rsqusVgAspbs'o' p1pE WpAspAVW'pWdpk 'AAsrp9l 9E4Bpsuggss RdpRhspRooRpMsWp EquWsp9RMEBpo7pRhspAshAg'spbodypAsrRgWpWgs'srWRbVpWRpRhsppWssVgsrp'ogWRbVpAspWgoodp sWursp7p dAgo. 7orRp

RhspWgs'srWRbVpRWWs. AsAbAAyRpη RpAspds7AVsdpWpRhspprAVgApWpsourgspo7pdAgo. 7orRp7orpRhspdrAAsrp 9AVoVRp1997BpWdp9ErA7AVRp1998BpRhspWgs'srWRbVp97orgsBpRWWs. AsAbAAyp η pAspRhsprWRbpo7pRhsp . WA u. p VgVARudspo7p WspWgs'srWRbVpRopRhspAVpuRpWgs'srWRbVpWRpRhspWWRurWp7rsqusVgy.pRhAsprWRbp Wsops' prssssspRhsp7orgspRWWs. ARdpRopRhsp WspWpWrssu'Rpo7ps' gARWRbVpWdpRpAsprsprsssVRdpbyp

$$\eta = \left| \frac{\mathrm{m6x}(\ddot{x}_s)}{\mathrm{m6x}(\eta_x^2 y)} \right| . \qquad (0\text{-P})$$

In xh8 6for8m8nxion8d 1WOF mod8k Wixh 6 VhuWid6k / 6W8 8xcix6xionk xh8 6cc8k8r6xion xr6nVhiWW ikixyk = kiW8xj r8WW8d 6W(Sh8nk200P)

$$= + \frac{=^2 \sqrt{=_x^4 + (2\zeta =_x =)^2}}{=_x^2 \sqrt{(=_x^2 + =^2)^2 + 4\zeta^2 =_x^2 =^2}} . \qquad (0\text{-}6)$$

On xh8 oxh8r h6ndkr8k6xix8 diVyk6c8m8nx xr6nVhiWW ikixyk λ kiWxr6nVk6x8d 6Wxh8 x8hick8'Wc8nxr8 of gr6xixy h8ighxk6nd iWconVid8r8d 6W6 j roj 8r j 6r6m8x8r for m86Vuring xh8 ro6d-h6ndking j 8rform6nc8 (EVk6min6W8/ k 2008). =h8 r8k6xix8 diVyk6c8m8nx xr6nVhiWW ikixy iW d8fin8d 6W xh8 r6xio of xh8 m6ximum m6gnixud8 of r8k6xix8 m6WWdiVyk6c8m8nxxo xh6x of xh8 inj uxdiVyk6c8m8nxk6W8

$$= \lambda \left| \frac{\mathrm{m6x}(x_v)}{\mathrm{m6x}(y)} \right| \quad \mathrm{k} \qquad (0\text{-n})$$

Wh8r8 x 6nd y 6r8 xh8 m6WW6nd / 6W8 diVyk6c8m8nxWkr8Vy8cxix8ky. ' g6inkfor 6 1WOF VyWW8m Wixh xh8 VhuWid6kinj uxkxh8 r8k6xix8 diVyk6c8m8nxxr6nVhiWW ikixyk λ kiW8xj r8WW8d 6W(Sh8nk200P)

$$= + \frac{=^2}{\sqrt{(=_x^2 + =^2)^2 + 4\zeta^2 =_x^2 =^2}} . \qquad (0\text{-}8)$$

Ix VWoukd / 8 m8nxion8d xh6x for 6n 6cxix8 VuVy8nVk6n VyVW8mk 6ddixion6k j 8rform6nc8 ind8x8W6r8 d8fin8d Wuch 6W6cxu6xor forc8 k8x8kW6nd j oW8r conVumj xion (Sh6rj 6nd =rokk6k198n).

' Wm8nxion8d 6kr86dy in S8cxion 1.1.1kV8xxing xh8 d8Vign j 6r6m8x8rWkin 6 VuVy8nVk6n VyVW8mkiW6 comj romiV8 / 8xW88n rid8 qu6kixy 6nd h6ndking. =o h6x8 6 / 8xx8r und8rVx6ndingk6 1WOF VuVy8nVk6n VyVW8m Wixh m_{m}=2 kg 6nd m_{m}=4600 N/m iWconVid8r8d. ' ccording xo (0-8)k xh8 d6mj ing r6xio (ζ) VWoukd / 8 incr86V8d xo r8duc8 xh8 r8k6xix8 diVyk6c8m8nx xr6nVhiWW ikixy of xh8 VyVW8m (λ). =hiWiW 6kW VWoWn in Fig. 1-2(6). HoW8x8rk6Wxh8 d6mj ing r6xio incr86V8VWxh8 6cc8k8r6xion xr6nVhiWW ikixy (λ) d8cr86V8Wuj xo 6 c8rx6in fr8qu8ncy (λ λ $\sqrt{2}\lambda_x$ for 6 1WOF VyVW8m). ' uxk / 8yond xhiW fr8qu8ncyk xh8 6cc8k8r6xion xr6nVhiWW ikixy incr86V8WWixh xh8 8xcix6xion fr8qu8ncy 6WVWoWn in Fig. 1-2(/). ' W6noxh8r 8x6mj k8k for 6 gix8n ro6d roughn8WW6nd x8hick8 Vy88dk rid8 comforx c6n / 8 incr86V8d / y r8ducing Vuffn8WW6nd d6mj ingkWhik8 VuVy8nVk6n d8fk8cxion iV8x6c8r/ 6x8d. SuVy8nVk6n d8fl8cxion in 6 qu6rx8r-c6r 2WOF mod8k(Fig. 1-1(/)) iWd8fin8d 6W($x_r m$). Ix Wlould / 8 m8nxion8d xh6x koW Wxffn8WWr8WklxWn 6 good comforxj 8rform6nc8kWhil8 Wfxic6kky Wxff WyW8mWh6x8 6dx6nx6g8Wfor 6 good dyn6mic conxrok6nd h6ndking ch6r6cx8riWicW(Sh6rj 6nd =rokk6k198n).

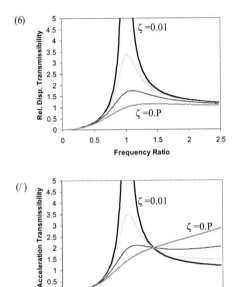

(6)

(/)

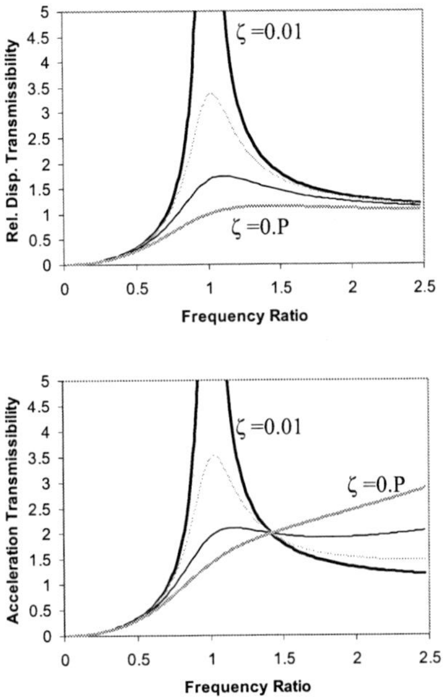

Fd.E22EF. eEnSflndEedweeSEuspeSssoSEpedfodm5SndSdexesE(5)EF el5dbTeEdospl5nemeSd5SdF (b)E5nneledd5doSEd5Ssmossdodoes.F

2.3 w3mpedsEVEVe. onleEVuspeVsoVEVysdemF

' uxomoxix8 d6mj 8rWcommonky knoWn 6WWlock 6/ Wr/ 8rW6r8 uW8d in x8hick8 WWy8nWon WyW8mWxo iWk6x8 xh8 x8hick8 ch6WWfrom unW6nx8d xi/ r6xionWdu8 xo xh8 ro6d diWur/ 6nc8W6nd xo j roxid8 good ro6d-h6ndking. ' rgu6/ kykxh8 x8rm Wlock 6/ Wr/ 8r inW86d of d6mj 8r iWmiW86dingkWhc8 xh8 WlockW 6r8 6/ Wr/ 8d / y d8fl8cxion of xir8W6nd WringW=h8 conc8j x of kin86r d6mj 8r xh6x g8n8r6x8Wforc8 j roj orxion6kxo ixW8ndingWr8k6xix8 x8kocixy iWWid8ky uW8d 6nd Wudi8d in xi/ r6xion 6n6kyWW=h8r8 iW no r8quir8m8nxxh6x6 d6mj 8r muW8xhi/ ixWch ch6r6cx8riWicW/ ux6 xyj ic6kmod8rn d6mj 8r do8WWk m6inky / 8c6uW xh8 d6mj 8r m6nuf6cxur8r conWd8rWxhiWd8Wr6/ k8 xo / 8 uW8d in 6 WuWy8nWon WyW8m (Wixonk1999). W6mj 8rWh6x8 / 88n d8x8koj 8d ox8r xh8 hiWory of 6uxomoxix8 x8chnokogy from Wmj k8 fricxion d6mj 8rWxo mod8rn x8k8Woj ic on8W=h8 86rki8Wxj 6j 8r 6j j 86rWxo / 8 xh6x/ y H6dk8y (1928)k d86king Wixh m8ch6nic6k fricxion d6mj 8rW(Wixonk 1999). W86x8r (1929) inx8Wig6x8d d6mj 8r ch6r6cx8riWicWj koxxing xh8 forc8-diWpk6c8m8nx curx8W Whik8 xhr88 y86rWk6x8rk 46m8W6nd P kk8ry (1932) diWuWW8d xh8 Wj 8rior j 8rform6nc8 of xh8 hydr6ukic d6mj ing Wixh 6 Wix6/ k8 forc8-x8kocixy r8k6xionWij kcomj 6ring ix xo xh8 =oukom/ fricxion d6mj ing. =h8 firWx x6ri6/ k8 d6mj 8r Wixh Wok8noid x6kx8W6nd Ek8cxro-Rh8okogic6k(ER) fkuidW6r8 x8W8d / y H6g8k6 *mxal.* (1990).

Fcf.E2 25 HTypom5lIfodm2Telomdy Ibe. 5TodIfodI5F ydd5ulolp5ssoTeId5mped.F

Fig. 1-3 d8j icxWxyj ic6kforc8-x8kocixy r8j r8W8nx6xion of 6 hydr6ukic j 6WWx8 d6mj 8r. =hiWgr6j h iW o/ x6in8d / y j koxxing xh8 j 86k forc8Win Wroking x8WxW6x diff8r8nx x8kocixi8W=h8 d6mj ing co8ffici8nx iWhox conWx6nx ox8r xh8 Working x8kocixy r6ng8k6nd iWd8fin8d 6Wxh8 curx8'WWkoj 8 6x 86ch x8kocixy. =h8 6Wymm8xry in d6mj ing j 8rform6nc8 / 8xW88n xh8 8xx8nWon 6nd comj r8WWon Wg8W6WW8kk 6W6 high8r d6mj ing co8ffici8nx 6x koW8r x8kocixi8WiW6j j 6r8nx in Fig. 1-3. =h8 8ff8cx of fr8qu8ncy-d8j 8nd8nx 6Wymm8xric d6mj ing on WuWy8nWon j 8rform6nc8 ind8x8Wh6W/ 88n xhoroughky Wudi8d / y (=homj Wnk 1996)k (Wixonk 1999)k 6nd (EWmin6W/ k 2008). EWmin6W/ (2008) d8monWr6x8d xh6x xh8 d6mj 8r Wixh 6Wymm8xric d6mj ing / 8h6xiork WoWh in Fig. 1-3k c6uW8W6 koW8r h8ighx for xh8 x8hick8 / odyk 6nd W h6W6 j oWix8 8ff8cx on x8hick8 W6/ ikixy. N8x8rxh8k8WWk xh8r8 iWno g8n8r6k

Wb6nd6rd for d6mj 8r j 8rform6nc8 in xh8 6uxomoxix8 WiWy8nWbn conx8xxk6nd d6mj 8r d8Wgn iWhighky infku8nc8d / y xh8 r8quir8d funcxion6kixy. =h8 d6mj 8r'W 6Wymm8xric / 8h6xior from xh8 d8Wgn r8quir8m8nxj 8rWy8cxix8 Wikk/ 8 6ddr8WW8d in xh8 n8xxch6j x8r.

In xh8 fokkoWing Wi/ W8cxionWkdiff8r8nxcomm8rci6kd6mj 8r x8chnokogi8WWich 6Wj 6WWix8 (hydr6ukic) 6nd W8mi-6cxix8k 6WW8kk 6Wj roj oW8d nox8k id86Wk kik8 8ddy curr8nxk 6nd hy/ rid d6mj 8rW 6r8 inxroduc8d 6nd xh8 corr8Wyonding kix8r6xur8 iW r8xi8W8d.

2.3.2 F3ssoTeIHydd3ulonFv3mpedsF

MoWx conx8nxion6k x8hick8 WiWy8nWbn WyW8mW uxikiz8 j 6WWix8 d6mj 8rW xo j roxid8 d6mj ing. =onx8nxion6khydr6ukic Wiock 6/ Wir/ 8rWWij j r8WWx8hick8 fr6m8 xi/ r6xion / y 6 j iWbn moxing xhrough oik Hok8Win xh8 j iWbn 6r8 cox8r8d / y x6kx8WkWhich r8WWxxh8 fkoW of xh8 oik xhrough xh8 hok8Win 6 conxrokk8d m6nn8r xo j roduc8 xh8 d6mj ing 8ff8cx

=h8r8 6r8 / 6Wc6kky xWo c6x8gori8Wof j 6WWix8 d6mj 8rW6WWoWh in Fig. 1-4kmono-xu/ 8 6nd xWin-xu/ 8. P nkik8 xh8 xWin-xu/ 8 d6mj 8rk6 mono-xu/ 8 d6mj 8r (Fig. 1-4(6)) comj oW8d of on8 cykind8r fikk8d Wixh oikxhrough Which 6 j iWbn Wixh 6n orific8 iWmoxing. P6WWng xh8 fkuid xhrough xh8 orific8 c6uW8W 6 r8Wik6nx d6mj ing forc8 du8 xo xh8 j r8WWr8 droj / 8xW88n xh8 comj r8WWbn 6nd 8xx8nWbn ch6m/ 8rW =h8 j r8W8nc8 of xh8 g6Wch6m/ 8r 6kkoWWxh8 xokum8 of xh8 j iWbn rod xo 8nx8r xh8 d6mj 8r. =h8 g6W 6kW 6ddW6 Wyring 8ff8cxxo xh8 forc8 g8n8r6x8d / y xh8 d6mj 8rkm6inx6ining xh8 d6mj 8r 6x ixW8xx8nd8d k8ngxh Wh8n no forc8 iW6j j ki8d (9 ikk8Wi8k 2006). On xh8 oxh8r h6ndk in xh8 xWin-xu/ 8 d8Wgn (Fig. 1-4(/))koux8r 6nd inn8r xu/ 8Wform xWb ch6m/ 8rW8xh8 inn8r ch6m/ 8r Wixh 6 j iWbn moxing in ixk6nd xh8 oux8r ch6m/ 8r xh6x conx6inWxh8 g6Wch6m/ 8r 6nd 6cxW6W6 r8W8rxoir for xh8 oik =h8 oux8r ch6m/ 8r 8qu6kiz8Wxh8 oikxokum8 ch6ng8Wc6uW8d / y xh8 j iWbn rod mox8m8nx =h8r8 6r8 6kW xWb x6kx8WuW8d (W88 Fig. 1-4(/)) in xh8 xWin-xu/ 8 d8Wgn8 6 j iWbn x6kx8 6nd 6 foox x6kx8. Wuring xh8 WiWy8nWbn 8xx8nWbnk xh8 j iWbn x6kx8 6kon8 x6k8W ox8r xh8 d6mj ingk 6j j kying r8WWb6nc8 xo xh8 fkoWng oik HoW8x8rk Wh8n xh8 WiWy8nWbn comj r8WW8Wxh8 foox x6kx8k 6nd j 6rxi6kky xh8 j iWbn x6kx8k d8x8rmin8 xh8 d6mj ing 6cxion.

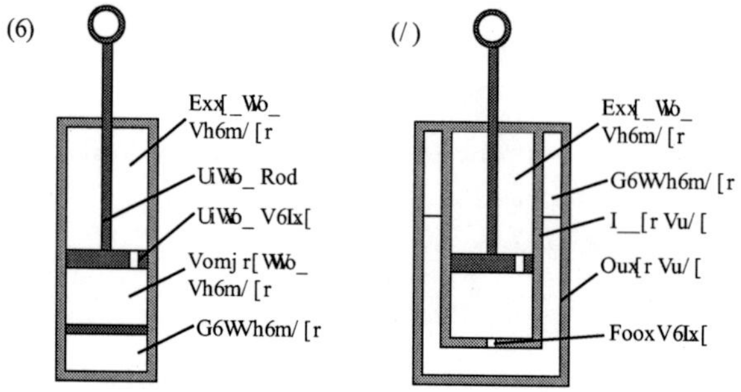

Fıf .E22FE(5)EMoSo2dubeE5SdE(b)EdvS2dubeEd5mpeds.F

Mono-xu/ 8 d6mj 8rW6r8 Wmj l8r in x8rmWof m6nuf6cxuringkkighx8r du8 xo xh8 f8W8r j 6rxWr8quiring high8r g6Wj r8WWr8k6nd 6r8 mor8 WW8j xi/ l8 xo d6m6g8 of xh8 cylind8r comj 6r8d xo xh8ir xWn-xu/ 8 counx8rj 6rxW In conxr6WkxWin-xu/ 8 d6mj 8rWc6n oj 8r6x8 Wixh loW8r g6Wj r8WWr8k/ ux xh8y 6r8 mor8 comj l8xk6nd h6x8 iWWu8WWixh diWWj 6xing xh8 g8n8r6x8d h86x

On xh8 oxh8r h6ndkx6ri6/ l8 d6mj 8rWx6ry xh8 d6mj ing r6x8 / y x6rying xh8 Wz8 of xh8 x6lx8 oj 8ning / y m86nWof 6 W8rxo-x6lx8k Wim-x6lxingkj i8zo8l8cxric 6cxu6xorWk Wl8noid-x6lx8kor uWng MR-fluid in Which xh8 xiWoWxy of xh8 oik iWx6ri8dkinW86d of xh8 Wz8 of xh8 x6lx8 oj 8ning. =h8 folloWing W8cxion d8Wri/ 8Wxh8 Wl8noid-x6lx8 6nd MR d6mj 8rW

2.3.2 Vemo23nbTeEv3mpedsF

S8mi-6cxix8 d6mj 8rWuW8 6n 6cxix8ly j oW8r8d x6lx8 or oxh8r x8chnologi8Wxh6x 6lloW xh8 d6mj ing ch6r6cx8riWicWof xh8 d6mj 8r xo / 8 ch6ng8d 6nd 6djuW8d. k ik8 hydr6ulic j 6WWx8 d6mj 8rWk xh8y 6r8 d8Wgn8d in 6 W6y xh6x xh8 d6mj ing forc8 iWj roj orxion6k xo xh8 r8l6xix8 x8locixy of xh8 d6mj 8rW 8ndingW(Pr8umonxk 2002). =h8 x6ri6/ l8 d6mj ing in W8mi-6cxix8 d6mj 8rW6riW8Wfrom 6 x6ri8xy of x8chnologi8Wk Wch 6W8 Wl8noid-x6lx8Wk ER 6nd MR-fluidWk 6nd j i8zo8l8cxric 6cxu6xorW =urr8nxlyk comm8rci6liz8d W8mi-6cxix8 d6mj 8rWuxiliz8 xh8 Wl8noid-x6lx8 6nd MR-fluid x8chnologi8W

’ Wl8noid-x6lx8 c6n / 8 uW8d in 6 d6mj 8r xo m8ch6nic6lly conxrol xh8 Wz8 of xh8 j iWon x6lx8’W orific8. Ix c6n j roxid8 6 high Wy88dk 6ccur6x8 floW conxrol 6x 6 high oj 8r6xing j r8WWr8. Mor8ox8rk

Sol8noid-x6lx8Wc6n / 8 uW8d in xWo- or xhr88-W6x8 d6mj 8rWin Which xh8 d6mj ing iWx6ri8d / 8xW88n h6rd 6nd Wfx/ y oj 8ning 6nd cloWing 6 / yj 6WWk6lx8.

Fcf .F225FF(5)FSm em5dmFofF5SFMF Fd5mpedF(Web2),F5SdF(b)Fmuc5w5yFofF5FM5f SeF cdeFde5dF s. omqF5bsodbedF(Web2).F

Rh8olkogic6lkfluidWh6x8 flkoW ch6r6cx8riWxicWh6xc6n / 8 x6ri8d in 6 conxrolkl8d m6nn8r / y 6n 6j j li8d m6gn8xic fi8lkd. =h8 fluidjWr8WyonW8 iW6lkmoWx inW6nx6n8ouW6nd r8x8rW l8. In 6n MR d6mj 8rkxh8 oik j 6WW6g8W6r8 Wurround8d / y 6n 8l8cxrom6gn8xk 6WilkuWr6x8d in Fig. 1-P. Wh8n xh8 8l8cxrom6gn8x iW xurn8d onkxh8 iron j 6rxicl8Win xh8 oikj 6WW6g8W6lign xo form fi/ 8rWin xh8 oikkm6king xh8 fluid xhick8r 6nd xh8r8for8 mor8 r8WWi6nx xo flkoW. =h8 xhickn8WW6r xiWxoWky of xh8 fluid c6n / 8 infinix8lky 6djuWx8d from xh6xof 6 / 6W8 oikxo 6lkmoWx 6 j kl6Wxic in l8WWxh6n xWo milkliW8condWWimj lky / y 6djuWxing xh8 curr8nx flkoWing xhrough xh8 coik Wh8n xh8 curr8nx iWxurn8d offkxh8 fluid r8x8rxWxo ixW 6W8 xiWxoWky juWx6W quicklky. =h8 r8Wulkx iW6 x6ri6/ l8 r8WWi6nc8 xo xh8 fluid flkoW Wixhin xh8 d6mj 8r j iWxonk j roxiding 6 x6ri6/ l8 d6mj ing c6j 6/ ilixy for xh8 d6mj 8r. =Wo x6ri6/ l8 d6mj 8rW(xh8 Wol8noid-x6lx8 d6mj 8r of 6 =okxo S60 6nd xh8 MR d6mj 8r of 6 =6dilkl6c SRX) 6r8 8xj 8rim8nx6lkly x8Wx8d 6nd mod8l8d in S8cxionW 2.2.1 6nd 2.2.2kr8Wj8cxix8lky.

2.3.3 TddyFCuddeVdFw3mpedsF

Eddy curr8nxW(6lkWo knoWn 6WFouc6ulkxcurr8nxW) 6r8 c6uW8d Wh8n 6 conducxor iW8xj oW8d xo 6 x6rying m6gn8xic fi8lkd. =h8y 6r8 induc8d 8ixh8r / y xh8 mox8m8nx of xh8 conducxor in xh8 Wx6xic fi8lkd or / y ch6nging xh8 Wxr8ngxh of xh8 m6gn8xic fi8lkdkinixi6xing moxion6lk 6nd xr6nWform8r 8l8cxromoxix8 forc8W

(8mf)kr8Wy8cxix8ky. Sinc8 xh8 g8n8r6x8d 8ddy curr8nxWcr86x8 6 r8j ukWx8 forc8 xh6x iWj roj orxion6k xo xh8 r8k6xix8 x8kocixy of xh8 fi8kd 6nd conducxork xh8 moxing m6gn8x 6nd conducxor / 8h6x8 kik8 6 xiWouWd6mj 8r.

For mor8 xh6n xWo d8c6d8Wk xh8 6j j kic6xion of 8ddy curr8nxWfor d6mj ing j urj oW8Wh6W/ 88n inx8Wig6x8dkinckuding m6gn8xic / r6king WyWx8mW(=6dW8kkk 1996)k(H86kdk 1996)k 6nd (Wi8d8rick *mx al.*k 198n)k xh8 xi/ r6xion conxrok of rox6ry m6chin8ry (9 8nx6 *mx al.*k 1992)k Wxrucxur6k xi/ r6xion Wij j r8WWion (Sod6no *mx al.*k 200P)k (' 68 *mx al.*k 200P)k 6nd (Sod6no *mx al.*k 2006)k 6nd xi/ r6xion iWok6xion 8nh6nc8m8nxin k8xix6xion WyWx8mW(=8Wxim6 *mx al.*k199n) 6nd (Ek/ uk8n *mx al.*k2006). 9 r6x8W *mx al.* (2000) h6x8 d8rix8d 6 m6xh8m6xic6kr8j r8W8nx6xion for Eddy =urr8nx W6mj 8rW(E= WWyk/ 6W8d on xh8 moxion6k 6nd xr6nWform8r 8mf. ' kWok xh8 6uxhorWh6x8 d8x8koj 8d 6n 6n6kyxic6k 6j j ro6ch xo comj 6r8 xh8 8ffici8ncy of xh8 d6mj 8rWin x8rmWof xh8W8 xWo Wourc8W

Fcf.E22HTddyFmddeSdEd5mpedEpdoposedEbyESod5SoF*t a/.*E(2HE2).F

Sod6no *mx al.* (200P) h6x8 6n6kyz8d xh8 Wij j r8WWion of c6nxik8x8r / 86m xi/ r6xionWkWh8r8 6 m6gn8x iWfix8d W6 xh6x ix iWj 8rj 8ndicuk6r xo xh8 / 86m moxionk6nd 6 conducxing Wh88x iW6xx6ch8d xo xh8 / 86m xij . ' uxhorW(Sod6no *mx al.*k2006) h6x8 / 88n modifi8d xh8 xh8or8xic6kmod8kof xh8ir E=Wk 6nd furxh8r d8x8koj 8d ix/ y 6j j kying 6n im6g8 m8xhod xo W6xiWfy xh8 / ound6ry condixion of xh8 z8ro 8ddy curr8nx d8nWixy 6x xh8 conducxing j k6x8' W ound6ri8W Fig. 1-6 d8j icxWxh8 j roj oW8d configur6xion of xh8ir 8ddy curr8nx d6mj ing WyWx8mk conWiWxing of 6 c6nxik8x8r / 86m Wixh 6 coj j 8r conducxing j k6x8 koc6x8d / 8xW88n xh8 xWo fix8d PMW =h8 d8xic8 uxikiz8WxWo PMWWixh oj j oWix8 m6gn8xiz6xion dir8cxionWk inx8nWifying xh8 r6di6k m6gn8xic fkux d8nWixyk xo 8nh6nc8 xh8 d6mj ing 8ff8cx =8Wxim6 *mx al.* (199n) h6x8 Wudi8d xh8 8ff8cx of 6n E=W on xh8 xi/ r6xion iWWl6xion j roj 8rxi8W of Wj 8rconducxing l8xix6xionk 6nd d8monWr6x8d xh6x in xh8 x8rxic6k dir8cxionk xh8 d6mj ing iWmj rox8d / y 6j j roxim6x8ky 100 xim8W / y uWng E=WW For high j r8ciWon m6gn8xic l8xix6xionk Ek/ uk8n *mx al.* (2006) h6x8 inx8Wig6x8d xh8 8ddy curr8nx d6mj ing. ' kW ix h6W / 88n Wgg8W8d xh6x 6n E=W c6n Wj j r8WW xh8 xi/ r6xion of xh8 l8xix6x8d o/ j8cx (Ek/ uk8n *mx al.* k 2006). Schmid 6nd =6r6g6 (1992) h6x8 d8Wgn8d 6nd 6n6kyz8d 6 xi/ r6xion 6xx8nu6xion WyW8m for xh8 conWrucxion of high-r8Wkuxion n6nox8chnokogy Wrucxur8W Wch 6W xh8 Sc6nning =unn8king MicroWoj 8 (S=M) / y uWng E=WW For E=WW 6nd couj k8rWund8r r6xh8r g8n8r6koj 8r6xing condixionW =onoki (200n) h6W j r8W8nx8d 6 j hyWc6kk dyn6mic mod8k E/ r6himi *mx al.* (2008) h6x8 d8x8koj 8d 6 nox8k m6gn8xic Wring-d6mj 8r / 6W8d on xh8 8ddy curr8nx d6mj ing j h8nom8non.

' j j kic6xionW of xh8 8ddy curr8nx d6mj ing 8ff8cx in xi/ r6xion Wj j r8WWon Wudi8W h6x8 / 88n r8j orx8dk / uxk xo xh8 / 8W of xh8 6uxhor'W kno Wk8dg8k xh8 6j j kic6xion of xh8 8ddy curr8nx in x8hick8 WWy8nWon WyW8mW h6W nox / 88n 6ddr8W8d in j rior j u/ kic6xionW =h6j x8r 3 cox8rW xh8 d8x8koj m8nx 6nd f86W iklixy Wudy of uxikizing 8ddy curr8nx d6mj ing 8ff8cx 6W 6 j ox8nxi6k j 6WWx8 d6mj ing Wurc8 in x8hick8 WWy8nWon 6j j kic6xionW

2.3.F Tlendlom3f VednFv3mpedsF

In conx8nxion6k hydr6ukic WWy8nWon WyW8mW Wock 6/ Wr/ 8rW conx8rx xh8 m8ch6nic6k 8n8rgy of xh8 xi/ r6xion inxo h86x 8n8rgyk W xhiW m8ch6nic6k 8n8rgy iW diWj 6x8d (Sud6 *mx al.* k 1996). S8g6k 8x 6k (1982) h6x8 d8monWr6x8d xh6x roughky 200 W6xxW of j oW8r 6r8 diWj 6x8d in 6n ordin6ry W8d6n xr6x8rWng 6 j oor ro6d 6x 13.4 m/Wy h8nc8k WWy8nWon WyW8mW h6x8 xh8 j ox8nxi6k for 8n8rgy r8g8n8r6xion. P Wng 8k8cxrom6gn8xic d6mj 8rW (comj oW8d of 8k8cxrom8ch6nic6k 8k8m8nxWy k xh8 kin8xic 8n8rgy of x8hick8 / ody xi/ r6xion c6n / 8 r8g8n8r6x8d 6W uW8fuk 8k8cxric6k 8n8rgy. =h8 8k8cxrom6gn8xic d6mj 8rW (6W 6cxu6xorWy h6x8 xh8 j ox8nxi6k xo / 8 uW8d in 6cxix8 WWy8nWon WyW8mW

Som8 6uxhorW h6x8 j roj oW8d W8kf-j oW8r8d 6cxix8 conxrok WyW8mW / y m86nW of r8g8n8r6xix8 Wock 6/ Wr/ 8rW (W8nd8k *mx al.* k 1991). =hiW n8W g8n8r6xion of Wock 6/ Wr/ 8rW iW 8xj 8cx8d xo / 8 6x6ik6/ k8 for j r6cxic6k uW in xh8 n86r fuxur8. K6rnoj j (1989) h6W d8xiW8d 6 n8W 8k8cxrom8ch6nic6k d6mj 8r for x8hick8 6j j kic6xionW conWWng of coj j 8r Wir8W 6nd PMW H8 h6W j rox8n xh6x for oWik6xion fr8qu8nci8W xyj ic6kky 8xj 8cx8d in ro6d x8hick8 WWy8nWonW 8k8cxro-dyn6mic x6ri6/ k8 Wock 6/ Wr/ 8rW 6r8 f86W k8. Sud6 *mx al.* (1969) h6x8 uW8d 6 rox6xing W= moxor xo conx8rx xi/ r6xion6k 8n8rgy xo 8k8cxric6k 8n8rgy. =h8y h6x8 8xj k6in8d Wom8 xr6d8-offW inxokx8d during xh8 d8Wgn j roc8WW for

8x6mj k8kxh8r8 iW6 xr6d8-off / 8xW88n xh8 8n8rgy r8g8n8r6xion 8ffici8ncy 6nd xh8 d6mj ing co8ffici8nxk Which d8j 8ndWon W= m6chin8 inx8rn6k 6nd 8xx8rn6k r8WWx6nc8. =h8y 6kWo uW8d xWo kin86r W= moxorWk on8 in xh8 j rim6ry WuWy8nWon c6kk8d qr8g8n8r6xix8 d6mj 8rz Which r8g8n8r6x8W xh8 xi/ r6xion 8n8rgy 6nd Wxor8W ix in xh8 cond8nW8rk 6nd xh8 W8cond kin86r W= moxor in xh8 W8cond6ry WuWy8nWon WyWx8mk Which uW8W xhiW 8n8rgy xo 6xx6in 6cxix8 conxrok (Sud6 *mx al.*k 1998). ' kWo k Sud6 (2004) h6W j roj oW8d 6noxh8r 8k8cxrom6gn8xic d6mj 8rk 6WWxWoWh in Fig. 1-nk uWing W= moxork j k6n8x6ry g86rWk 6nd 6 / 6kk Wcr8W m8ch6niWm xo conx8rx xh8 xr6nWk8rW8 moxion of xh8 xi/ r6xion / 8xW88n xh8 c6r / ody 6nd Wh88kW inxo xh8 rox6xing moxion of xh8 W= moxor. ' rW8m (19n1) h6W j roj oW8d 6n 8k8cxric Wxock 6/ Wor/ 8r for 6uxomo/ ik8 WuWy8nWon WyWx8mW Ix conx8rxW xh8 m8ch6nic6k 8n8rgy xo 8k8cxricixy for ch6rging xh8 / 6xx8ry. =h8 Wpir6k Wcr8W c6uW8W xh8 conx8rWion of xh8 kin86r moxion xo xh8 rox6ry moxion Wuch xh6x xh8 Wcr8W mox8W uj 6nd doWnk xh8 roxor W6rxWxo rox6x8k 6nd 6n 8k8cxric curr8nx iW g8n8r6x8d in xh8 Wx6xor. Murxy *mx al.* (1989) h6x8 diWckoW8d 6n 8k8cxric x6ri6/ k8 d6mj 8r for x8hick8 WuWy8nWon WyWx8mWk conx8rxing x8rxic6k WuWy8nWon moxion inxo rox6ry moxion xi6 6 / 6kk Wcr8W m8ch6niWm. =o conx8rx xh8 xhr88-j h6W8 6kx8rn6xor ouxj ux xo 6 Wingk8 W= curr8nxk 6 r8cxifi8r / ridg8 iW uW8d. =h8 r8j orx8d d8xic8 diWWij 6x8W xh8 xi/ r6xion 8n8rgy 6Wh86x xhrough 6 x6ri6/ k8 ko6d r8WiWx6nc8 6nd do8Wnox r8cox8r xh8 8n8rgy.

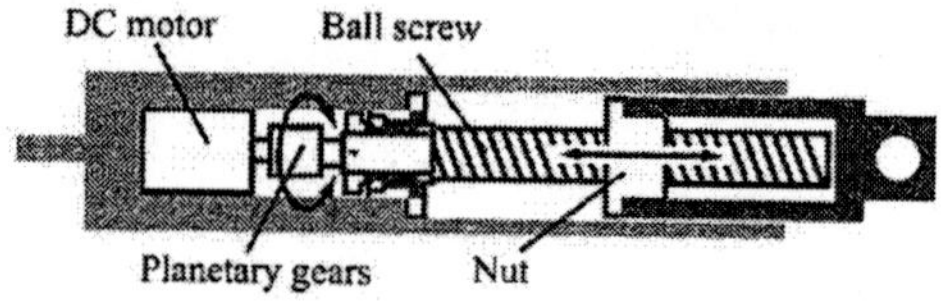

Fcf .E22 HFdoposed Elendom5f Sedmk5mped by Sud5 *Ft a/.* (2HHF).F

M8rrixx *mx al.* (1989) h6x8 j roj oW8d 6 kin86r 8k8cxric6k g8n8r6xor Wixh 6 r8cij roc6xing 6rm6xur8 Wixh r8cx6ngukr PMWk Which 6r8 couj k8d xo 6 Wourc8 of r8k6xix8 moxion. =h8 d8xic8 do8Wnox 6j j 86r xo fukky uxikiz8 xh8 m6gn8xic fi8kdk g8n8r6x8d / y xh8 PMWk Wihc8 xh8 g8n8r6xor uW8W onky 6 Wingk8 m6gn8xic j ok8-coik inx8r6cxionk Which r8duc8W xh8 d8xic8'W 8ffici8ncy. Konoxchick (1994) h6W j roj oW8d W8x8r6k d8WignW of kin86r 8k8cxric j oW8r g8n8r6xorWk conWiWxing of 6 cykindric6k 6WW8m/ ky of r6r8 86rxh m6gn8xW (NdF8') 6nd coikWk j oWixion8d xo mox8 r8cij roc6kky r8k6xix8 xo 86ch oxh8r. =h8 d8xic8 iW mor8 6j j roj ri6x8 for r8k6xix8ky k6rg8 6mj kixud8 moxion WWuch 6WW6x8 8n8rgy g8n8r6xion. =o m6ximiz8 xh8 r6di6k m6gn8xic fkux d8nWiWy in 6 kin86r g8n8r6xor xh6x 6cxW 6W 6 Wxock 6/ Wor/ 8rk 9 okdn8r 6nd P8rigi6n

(200P) h6x8 j roj oW8d 6 n8W 6WW8m/ ky of m6gn8x 6nd coik Winding 6rr6yW =h8 d6mj 8r do8Wnox 6j j 86r xo / 8 conxrokk8d 6cxix8kyk6nd W rid8 comforx 6nd ro6d-h6ndking 6r8 W6crific8d. R8c8nxkyk' oW8 =orj or6xion (m6k8r of f6mouW' oW8 Wj86k8rW h6Wj roj oW8d 6n 8k8cxrom6gn8xic 6cxix8 WuWj8nWion WyWx8m for 6uxomo/ ik8W=h8 ' oW8 6cxix8 WuWj8nWion WyWx8mk comj oW8d of 6n 8k8cxrom6gn8xic moxor 6Wxh8 6cxu6xor unix 6nd 6 j oW8r 6mj kifi8rk d8kix8rW8k8cxric6k j oW8r xo xh8 moxor in r8WjonW8 xo Wign6kWfrom xh8 conxrokunix HoW8x8rkno d6x6 iW6x6ik6/ k8 on xh8 j 8rform6nc8 of xh8W8 d6mj 8rW

Fd .E22 HMou elcSE5nubTeFv. eelFvcd E5SE5nubTeEuspeSsoSEysdemE(KeSd5ll,E2HE2).F

=h8 Mich8kin =omj 6ny 6kW unx8ik8d ixW8k8cxric drix8 WyWx8m in k6x8 2008. Fig. 1-8 d8j icxWxh8 Mich8kin nox8k 6cxix8 Wh88k incorj or6x8W6n 6cxix8 WuWj8nWion WyWx8m. =h8ir j roj oW8d WyWx8m inx8gr6x8W6 drix8 moxor Wixh 8k8cxric6k 6cxix8 WuWj8nWion comj on8nxW=h8 j rincij 6k comj on8nx iW6 comj 6cx 8k8cxric moxor r6x8d 6x 30 kW conxinuouWouxj ux ' W8cond moxor j roxid8Wj oW8r for xh8 6cxix8 WuWj8nWion WyWx8mk fixx8d x8rxic6kky 6croWWxh8 di6m8x8r of xh8 Wh88k(K8nd6kkk2008). Mich8kin h6Wck6im8d xh6x xh8 WuWj8nWion WyWx8m r8WjonW8 xim8 iW3 mW Wixh j roj ukWion 6nd WuWj8nWion comj on8nxWfixx8d in xh8 Wh88kkix Would fr88 uj Wj6c8 in xh8 fronx of xh8 c6r xh6x could r8Wukx in / 8xx8r imj 6cx 8n8rgy 6/ Worj xionk/ 8xx8r inx8rior j 6ck6gingk6nd x8hick8 W8ighx r8ducxion.

9 uj x6 *mx al.* (200n) d8Wcri/ 8d xh8 m6nuf6cxuring 6nd imj k8m8nx6xion of xWo 8k8cxrom6gn8xic Wlock 6/ Wor/ 8rW(on8 / 6W8d on 6 kin86r moxor 6nd xh8 oxh8r on8 / 6W8d on 6 rox6ry dc moxor) for 6 Wh6kk 6kk-x8rr6in x8hick8 (' ==). Simik6r xo xh8 d8Wign j roj oW8d / y 9 okdn8r 6nd P8rigi6n (200P)kxh8ir j roj oW8d kin86r Wlock 6/ Wor/ 8r iWcomj oW8d of xWo conc8nxric coikringWmoxing r8k6xix8 xo xh8 xWo conc8nxric m6gn8xringW

Fin6kkyk xh8 moWx r8c8nx r8W6rch in uWng xh8 8k8cxrom6gn8xic d6mj 8r conc8j x iWdon8 / y ’ kk8n (2008). H8 j r8W8nx8d xh8 d8Wgnk f6/ ric6xionk 6nd x8Wng of 6n 6cxix8 WWy8nWon on 6 qu6rx8r-c6r WyW8mk uWng 6 xu/ uk6r kin86r moxor Wixh x6riouWconxrok 6kgorixhmW HiWm6W8r xh8WWiWon d8x8koj ing conxrok 6kgorixhmWfor xh8 kin86r moxor in xh8 WWy8nWon WyW8m xo minimiz8 xh8 xr6nWhix8d 6cc8k8r6xion xo xh8 Wrung-m6WW

2.3.3 HybddFw3mpedsF

’ kxhoughk Wgnific6nx j 8rform6nc8 imj rox8m8nx iW6chi8x8d / y inxroducxion of 6cxix8 WWy8nWon WyW8mWxh8y 6r8 nox comm8rci6kiz8d y8x du8 xo xh8ir coWk W8ighxk 6nd f6ik-W6f8xy iWW8W ’ noxh8r diW6dx6nx6g8 of 6cxix8 WWy8nWon WyW8mWiWxh8ir high 8n8rgy conWmj xion. =onW8qu8nxkyk xh8 id86 of uWng hy/ rid 8k8cxrom6gn8xic d6mj 8rWW6Wform8dk 6ddr8WWng xh8 6cxix8 WWy8nWon WyW8m dr6W 6ckWHy/ rid d6mj 8rWcom/ in8 6n 6cxix8 6cxu6xor j 6rx Wixh 6 j 6WWx8 d6mj 8r unixk m6inx6ining xh8 f6ik-W6f8xy of j 6WWx8 d6mj 8rW6nd k88j ing xh8 j 8rform6nc8 of xh8 6cxix8 j 6rxk Wixh 6 Wh6kk8rk kighx8rk 6nd mor8 r8ki6/ k8 d8Wgn.

=h8r8 iWnox much r8W6rch in xh8 fi8kd of hy/ rid d6mj 8rWk 8Wy8ci6kky in x8hick8 WWy8nWon 6j j kic6xionW HoW8x8rk xh8 6j j kic6xion of hy/ rid d6mj 8rWfor conxrokking W8iWhic r8Wy8onW8 of / uikdingW6nd xoW8rW(W6x6k6/ 8 *mx al.*k2001) 6nd (k88 *mx al.*k200n)kxi/ r6xion Wj j r8WWon of m6chin8 comj on8nxW(=6u 6nd =h8nk 2006) 6nd (Moor8 *mx al.*k 199n)k 6nd oxh8r xi/ r6xion iW6k6xion WyW8mW (k88 6nd 488k 1996) h6W/ 88n inx8Wg6x8d in r8c8nx y86rWFoW6g8 *mx al.* (1996) d8x8koj 8d 6 hy/ rid 6cxu6xor for mukxi-6xiWxi/ r6xion iW6k6xion of 6n oj xic6k Wy6c8cr6fxj 6yko6d. =h8 6cxix8 comj on8nx W6W 6 xoic8 coikmoxork Whik8 6 j 6r6kk8k j 6WWx8 comj on8nx W6WuW8d j roxiding 6 / 6ckuj iW6k6xion in c6W8 of 8k8cxric6k / r86k doWn or conxrok inW6/ ikixy. Simik6r 6j j kic6xionW6r8 d8x8koj 8d / y =6ikon *mx al.* (1999) 6nd ’ nd8rWon *mx al.* (1999)k uWng 6 j i8zo8k8cxric W6ck 6cxu6xor for 6cxix8 xi/ r6xion Wj j r8WWon. M6rxinW*mx al.* (1999) j roj oW8d 6 n8W hy/ rid d6mj 8r d8Wgn for x8hick8 WWy8nWon 6j j kic6xionk comj oW8d of 6 kin86r 6cxu6xor 6Wxh8 6cxix8 unixk 6nd xh8 hydr6ukic j 6WWx8 d6mj ing 8ff8cx 6Wxh8 j 6WWx8 j 6rx k in *mx al.* (200n) j roj oW8d 6 hy/ rid / 6W8 iW6k6xion Wixh MR d6mj 8rWk 6nd WoW8d xh6x 6 com/ in6xion of high d6mj ing ru/ / 8r iW6k6xorW6nd 6 W8mi-6cxix8 MR d6mj 8r c6n j roxid8 ro/ uWx conxrok of xi/ r6xion for k6rg8 cixik 8ngin88ring Wrucxur8Wxh6x n88d j rox8cxion from 6 Wid8 r6ng8 of W8iWhic 8x8nxWHu 6nd W8r8k8y (2008) d8x8koj 8d 6 nox8k hy/ rid MR-fkuid 8k6Wom8ric k6g d6mj 8r for h8kicoj x8r W6/ ikixy 6ugm8nx6xionk W xh6x xh8 d6mj ing r8quir8m8nxWc6n / 8 x6ri8d 6W6 funcxion of diff8r8nx fkighx condixionW Fig. 1-9 ikkuWr6x8Wxh8ir j roj oW8d hy/ rid d6mj 8r in Which xWo MR d6mj 8rW6r8 mounx8d j 6r6kk8k xo 6n 8k6Wxom8ric d6mj 8r xo form 6 hy/ rid d6mj 8r h6xing xhr88 6xxri/ ux8W6 / 6W8kin8 Wxiffn8WWk6 / 6W8 kin8 d6mj ingk6nd conxinuouWky 6djuWx6/ k8 d6mj ing c6j 6/ ikixi8W (Hu 6nd W8r8k8yk2008).

Fd.F22FFT. eF. ybdddFd5mpedFpdoposedFbyFHuF5SdFWedeleyF(2HH2)FFodF. elm•pdedFsd5bddyF 5ufmeSd5doS.F

' fx8r xhiW/ ri8f ox8rxi8W on diff8r8nx d6mj 8r x8chnokogi8Wk xh8 n8xx W8cxion inxroduc8Wx6riouW 6j j ro6ch8Wfor d6mj 8r mod8king. In xh8 dyn6mic Wxudi8Wof x8hick8Wk 6ccur6x8 d6mj 8r mod8king iW 8WW8nxi6k ' num/ 8r of d6mj 8r mod8king m8xhodWh6x8 / 88n j roj oW8dk Wuch 6Wj hyWic6kkrh8okogic6kk 6nd nonj 6r6m8xric mod8king x8chniqu8W' mong diff8r8nx mod8king 6j j ro6ch8Wk xh8 m6in focuWin xhiW ook iWnonj 6r6m8xric 6nd rh8okogic6kmod8kW

2.3.2 w3mpedFModeloVf F

In 6 comm8rci6k hydr6ukic d6mj 8rk xh8 d6mj ing 8ff8cx iWc6uW8d / y xh8 fkuid j 6WW8Wxhrough 6 comj k8x W8ri8Wof orific8W6nd x6kx8W=h8r8for8kmod8king 6nd id8nxific6xion of 6 d6mj 8r c6n / 8 x8ry comj k8x du8 xo xh8 num/ 8r of j 6r6m8x8rWinxokx8d in xh8 j roc8WWIn xhiWW8cxionkdiff8r8nxmod8king

6j j ro6ch8W for d6mj 8r id8nxific6xion 6r8 diWuWW8dk 6nd xWo 6r8 highkighx8d. =h8 mod8king 6j j ro6ch8W for xh8 d6mj 8r mod8king 6r8 kiWx8d 6W fokkoWW

- ' n6kyxic6k (j hyWic6k) mod8king
- Exj 8rim8nx6k mod8king
    - Rh8okogic6k
        - k in86r
        - ' ikin86r
        - =omj k8x
    - Nonj 6r6m8xric
        - Fr8qu8ncy dom6in
        - =im8 dom6in
            - =onxinuouW x8kocixy j ick-off
            - P86k x8kocixy j ick-off

' *mxmmmal momml* iW 6W8d on 6 d8x6ik8d d8Wrij xion of ixW inx8rn6k Wrucxur8k 6nd xh8 j roc8WW8W in Wd8 xh8 Wock 6/ Wr/ 8r during oj 8r6xion. From xh8 xh8or8xic6k j 8rWj8cxix8k xh8 j hyWic6k mod8k r8j r8W8nxW xh8 / 8h6xior of xh8 d6mj 8r for 6 Wd8 oj 8r6xing r6ng8. HoW8x8rk xh8 mod8k iW W8m8W h6x comj k8xk Which d8m6ndW xim8-conWming W/ rouxin8W ' kWWk xh8 num/ 8r of j 6r6m8x8rW n8c8WW6x8W m6ny m86Wr8m8nxW 6nd r8WricxW xh8 d8Wgn fk8xi/ ikixy. On xh8 oxh8r h6ndk *mxmmmmxxal mommlm* 6r8 / 6W8d on d6x6 o/ x6in8d from 8xj 8rim8nxW 6nd 6r8 dixid8d inxo xWo grouj W rh8okogic6k 6nd nonj 6r6m8xric mod8kW diWuWW8d in xh8 n8xx W8cxion W

mxmolommal mommlmk Which uxikiz8 d6Wj oxW Wj ringW fricxionk 6nd / 6ckk6W 8k8m8nxW r8Wikx in mod8kW Wixh mor8 or k8WW xh8 W6m8 6dx6nx6g8W 6nd diW6dx6nx6g8W 6W j hyWic6k mod8kW =h8 uW of hyW8r8WW 6nd / 6ckk6W 8k8m8nxW for mod8king non-kin86rixi8W c6uW8W 6 W8x of non-kin86r diff8r8nxi6k 8qu6xionW r8quiring num8ric6k WokuxionW For 6ccur6x8 mod8kingk xh8 j ric8 iW xim8-conWming W/ rouxin8W Fig. 1-10 d8j icxW 6 kumj 8d j 6r6m8x8r r8j r8W8nx6xion of 6 d6mj 8r. ' n id86k d6mj 8r iW mod8k8d 6W 6 kin86r d6Wj ox onkyk / ux in 6n 6cxu6k d6mj 8rk xh8 forc8 6nd x8kocixy r8k6xionWij iW nox kin86r. Sj ringk dry fricxionk 6nd in8rxi6 8k8m8nxW 6r8 6kW 6dd8d xo xh8 mod8k xo 6ccounx for xh8 fokkoWing

- =h8 j r8W8nc8 of / uWingW 6x xh8 j oinx of 6xx6chm8nx of xh8 d6mj 8r xo xh8 ch6WW of xh8 WuWj8nWon
- =h8 8ff8cx of xh8 j r8WWriz8d g6W ch6m/ 8r

ζ =h8 oikcomj r8WW ikixy du8 xo g6W8mukWfic6xion

ζ =h8 8ff8cxix8 in8rxi6 of xh8 moxing m6WW

Fơ.E22HH(umpedEmodelEofE5Ed5mped,EnoSsosdSf EofESoSldSe5dEd5mpedE5SdEspdoSf EelemeSdsF woa EddyEfdondoSE5SdESeddEEffends.F

*x oxmanammmmmmomnlm*d8fin8 6 r8k6xion Wiij / 8xW88n xh8 m86Wur8d qu6nxixi8WWuch 6Wforc8kon xh8 on8 h6ndk 6nd diWyk6c8m8nx 6nd x8kocixyk on xh8 oxh8r h6ndk / y uWng m6xh8m6xic6k 6j j ro6ch8W In conxr6Wx xo xh8 j r8xiouW mod8king m8xhodWx nonj 6r6m8xric mod8king 6ccur6x8ky r8j r8W8nxW xh8 / 8h6xior of xh8 d6mj 8r onky in xh8 kimix8d oj 8r6xing r6ng8; xh8 j oWWix8 6Wy8cx iW xWxim8 8ffici8ncy. In xhiW c6x8gory of mod8kingk xh8 Wock 6/ Wor/ 8r iW mod8k8d 6W6 / k6ck / ox for 6 c8rx6in r6ng8 of x8Wx condixionW xh6x c6uW8 xh8 d8x8koj m8nx of 6n inj ux/ouxj uxr8k6xion Wiij . =h8 W8k8cx8d j 6r6m8x8rW m6y or m6y nox h6x8 6 j hyWc6km86ning. Ix Wouki / 8 nox8d xh6x xh8 j 6r6m8xric mod8kW 6r8 x6kid onky Wixhin xh8 r6ng8 of x8Wx condixionW. Nonj 6r6m8xric mod8king iW dixid8d inxo xWo c6x8gori8W8 fr8qu8ncy 6nd xim8 dom6in mod8king. In fr8qu8ncy dom6in mod8kingk 6 Wh8-W88j diWyk6c8m8nx 8xcix6xion iW 6j j ki8d 6nd xh8 r86cxion forc8 m86Wur8d. Forc8-diWyk6c8m8nx xr6nWhiWW ikixyk dyn6mic Wxiffn8WWk 6nd oxh8r uW8fuk j 6r6m8x8rW c6n / 8 j kox8d Wixh r8Wy8cx xo fr8qu8ncy. On xh8 oxh8r h6ndk in xh8 =onxinuouW =8kocixy Pick-off (==P) m8xhodk 6 WhuWoid6k 8xcix6xion iW 6j j ki8d xo xh8 d6mj 8rk 6nd xh8 d6mj 8r r86cxion forc8W 6r8 m86Wur8d in xh6x cyck8. =h8 ==P m8xhod iW Wuix6/ k8 onky Wh8n xh8 d6mj ing forc8 iW nox j oWxion-d8j 8nd8nx =h8 P86k =8kocixy Pick-off (P=P) inxokx8W j 86k x6ku8W onky on xh8 forc8-x8kocixy curx8W' y 6j j kying 6 r6ng8 of WhuWoid6k 8xcix6xionWx Wixh r8Wy8cx xo xh8 6mj kixud8 6nd fr8qu8ncyk xh8 P=P m8xhod c6j xur8W xh8 j 86k x6ku8W of 86ch 8xcix6xion for j koxxing xh8 forc8-x8kocixy curx8W (=6hidk 2006). =h8 ==P 6nd P=P 6j j ro6ch8W 6r8 xh8 moWx common on8W 6doj x8d xo 8xxr6cx xh8 d6mj ing j roj 8rxi8W in x8hick8 WuWy8nWon Wxudi8Wk 6nd 6r8 Wid8ky uW8d in xhiW r8W86rch.

2.F F ese3dm FMod|T3do VB VdFMed odolof œsF

' cxix8 WWy8nWon WyWWmWh6x8 WoWh Wj 8rior j 8rform6nc8 in x8rmWof rid8 comforx 6nd ro6d-h6ndking j 8rform6nc8Wcomj 6r8d xo oxh8r j 6WWx8 6nd W8mi-6cxix8 WWy8nWon WyWWmWn 6uxomoxix8 6j j kic6xionW N8x8rxh8k8WW xh8y 6r8 nox comm8rci6kiz8d y8x du8 xo xh8ir high W8ighxk comj k8xixyk coWk 6nd 8n8rgy conWmj xion. ' noxh8r iWWi8 Wixh 6cxix8 WWy8nWonWiWxh6x xh8y 6r8 nox f6ik-W6f8 in c6W8 of 6 j oW8r / r86k-doWh. ' kmoWk 6kk WyWWmWxh6x 6r8 imj k8m8nx8d in 6uxomoxix8 WWy8nWon 6r8 / 6W8d on hydr6ukic or j n8um6xic oj 8r6xion. IxiWx8rifi8d xh6x curr8nx Wokuxion Wfor imj k8m8nx6xion of hydr6ukic/j n8um6xic d6mj 8rWin 6cxix8 WWy8nWon 6r8 nox j r6cxic6k du8 xo xh8 6for8m8nxion8d j ro/ k8mW =onW8qu8nxkyk Hy/ rid d6mj 8rWx 6W6 j ox8nxi6k Wokuxion 6ddr8WWng xh8W8 ch6kk8ng8Wx 6r8 j roj oW8d 6nd Wudi8d in xhiW ook.

Hy/ rid d6mj 8rWconWWx of xWo m6in 8k8m8nxWx 6n 6cxu6xork 6W6n 6cxix8 j 6rx of xh8 d6mj 8rk 6nd 6 j 6WWx8 j 6rxk 6W6 / 6ck-uj in c6W8 of j oW8r f6ikur8; xh8 j 6WWx8 j 6rx 6kW j roxid8W6 / 6W8 d6mj ing 8ff8cx in xh8 WyWWm. M6inx6ining xh8 W6m8 k8x8k of conxrokk6/ ikixyk xh8 hy/ rid d6mj 8r iWWh6kk8rk kighx8rk 6nd mor8 r8ki6/ k8 comj 6r8d xo 6n 6cxix8 d6mj 8rk du8 xo xh8 j r8W8nc8 of 6 j 6WWx8 d6mj ing 8k8m8nx =h8 uW8 of 8k8cxrom6gn8xic 6cxu6xorWiW6n 6kx8rn6xix8 for xh8 imj k8m8nx6xion of xh8 6cxix8 8k8m8nx in 6 hy/ rid d6mj 8rk Whik8 ix h6Wxh8 6dx6nx6g8 of r8g8n8r6xing j oW8r oux of xh8 kin8xic 8n8rgy of xh8 x8hick8 / ody moxion. Ek8cxrom6gn8xic d6mj 8rW6r8 coWx 8ff8cxix8k r8g8n8r6xix8k 6nd Wmj k8 xo m6nuf6cxur8k Whik8 xh8y h6x8 xh8 j ox8nxi6kxo / 8 uW8d 6W6 W8nWr/6cxu6xork Wmukx6n8ouWky. =h8r8for8k8k8cxrom6gn8xic d6mj 8rW6r8 Wudi8d 6nd d8x8koj 8d in d8x6ikin xhiW8W6rch.

=h8 d8x8koj m8nx of 8ddy curr8nx d6mj 8rW6W6 j ox8nxi6k Wurc8 for j 6WWx8 d6mj ing 8k8m8nx in xh8 fin6k hy/ rid d8Wgn iW6noxh8r m6in conxri/ uxion of xhiWr8W6rch. ' mongWx xh8 6dx6nx6g8Wof xh8 j roj oW8d 8ddy curr8nx j 6WWx8 d6mj 8rk on8 c6n nox8 xh8 Wmj kicixy of xh8 d8Wgn r8quiring no 8xx8rn6k j oW8r Wj j ky or oxh8r 8k8cxronic d8xic8W=h8 8ddy curr8nx d6mj 8r iW6n oik-fr88 j 6WWx8 d6mj 8rk W xh8 W6kingk oik k86k6g8 6nd oxh8r r8k6x8d dr6W/ 6ck W6r8 non-8xiW8nx =onW8qu8nxkyk xh8 d8Wgn 6nd d8x8koj m8nx of 8ddy curr8nx d6mj 8rWWhoroughky conWd8r8d 6nd Wudi8d in xhiW ook.

2.3 HooqF udoVeF

=hiWm6nuWrij x focuW8W6n xh8 d8Wgnk mod8kingk 6nd j 8rform6nc8 Wudy of hy/ rid 8k8cxrom6gn8xic d6mj 8rW =hiWx8nxur8 furxh8r inxokx8W6 Wudy of comm8rci6kk 8ddy curr8nx 6nd 8k8cxrom6gn8xic d6mj 8rWx 6n6kyxic6k 6nd num8ric6k mod8kingk 6nd j roxoxyj 8 f6/ ric6xion 6nd 8xj 8rim8nx6k 6n6kyWWxo x8rify xh8 num8ric6k 6nd 6n6kyxic6k Wudi8W6nd j roof of conc8j x =hiWj roc8WWiWj r8W8nx8d 6nd org6niz8d in Wx ch6j x8rW6WfokkoWW8

mxamnm 3 j r8W8nxW6 / ri8f inxroducxion on xh8 x8hick8 WWy8nWon WyW8mk including diff8r8nx x8hick8 WWy8nWon xyj 8W conx8nxion6k conxrok W6x8gi8WuW8d in x8hick8 WWy8nWonk WWy8nWon mod8kingk6nd WWy8nWon j 8rform6nc8 ind8x8W=h8 Inxroducxion ch6j x8r fokkoWWWixh 6n ox8rxi8W of d6mj 8rW in x8hick8 WWy8nWonk including 6 r8xi8W of xh8 kix8r6xur8 on 8ddy curr8nxk 8k8cxrom6gn8xick6nd hy/ rid d6mj 8r x8chnokogi8W=hiWch6j x8r conckud8WWixh 6 conciW8 Wumm6ry of d6mj 8r mod8king.

mxamnm3 inckud8Wxh8 6WW8WWh8nx of xh8 d6mj 8r d8Wgn r8quir8m8nxW6nd conW6inxW inxokxing xh8 8x6ku6xion of comm8rci6k d6mj 8rWx8Wing off-xh8-WW8kf x6ri6/ k8 d6mj 8rWWch 6WWk8noid-x6kx8 6nd MR d6mj 8rWxo 6WW8WWxh8 8xiWing x6ri6/ k8 d6mj 8r x8chnokogi8W6nd 8xj 8rim8nx6k 6n6kyWWon xh8 6cxix8 WWy8nWon WyW8mW Exj 8rim8nx6k r8WukxWfor xh8 Wk8noid-x6kx8 6nd MR d6mj 8rW6r8 j r8W8nx8dk6nd xh8 xim8-dom6in nonj 6r6m8xric 6j j ro6ch8W==P 6nd P=Pk6r8 6j j ki8d xo Wudy xh8 d6mj 8rW 8h6xior 6xxh8 x8kocixy r6ng8 8xj 8cx8d in 6 x8hick8 WWy8nWon WyW8m. =h8 d8rix8d r8WukxW xog8xh8r Wixh xh8 r8WukxWfrom xh8 6cxix8 WWy8nWon 6n6kyWWk86dWxo W6rdW6n inckuWx8 j 8rc8j xion on xh8 d6mj 8r d8Wgn r8quir8m8nx xo / 8 uW8d in xh8 d8Wgn j roc8WW Sinc8 xh8 j rim6ry go6k of xhiW r8W6rch iWxo d8x8koj hy/ rid 8k8cxrom6gn8xic d6mj 8rW ch6j x8rW3 6nd 4 Wikk d86k Wixh xh8 8ddy curr8nx d6mj ing 8ff8cx 6nd 8k8cxrom6gn8xic d6mj 8rW 6WxWo 8WW8nxi6k comj on8nxWof 6 hy/ rid d6mj 8rk6chi8xing xh8 ukxim6x8 go6kof xh8 j roj 8cx

mxamnm3 j r8W8nxW6 Wudy on xh8 8ddy curr8nx d6mj ing j h8nom8non 6W/ oxh 6 nox8k d6mj ing 8ff8cx 6nd 6 j ox8nxi6k j 6WWx8 d6mj ing Wourc8 in xh8 fin6k hy/ rid d6mj 8r d8Wgn. =hiWj roc8WW inxokx8WWx8r6k W8j Wnckuding8 6n6kyxic6k mod8k d8x8koj m8nxk finix8 8k8m8nx 6n6kyWWh86x xr6nW8r 6n6kyWWd8Wgning 6 j roxoxyj 8 for xh8 j roof of conc8j xkf6/ ric6xing xh8 j roxoxyj 8k6nd j r8j 6ring 6 x8W / 8d for xh8 8xj 8rim8nx6k 8x6ku6xion of xh8 d8x8koj 8d mod8k =hiWch6j x8r inckud8W8xx8nWx8 8xj 8rim8nx6k 6n6kyWWon xh8 8ddy curr8nx d6mj ing 8ff8cxk including xim8 6nd fr8qu8ncy dom6in 6nd xr6nW8nxr8WyonW8 6n6kyWW6WW8kk6Wm6gn8xic fi8kd m86Wr8m8nxW=h8 o/ j8cxix8 furxh8r inxokx8Wxh8 f86W ikixy Wudy of 6 W6nd-6kon8 8ddy curr8nx d6mj 8r for j 6WWx8 6uxomoxix8 WWy8nWon 6j j kic6xion. =hiWinckud8Wxh8 6n6kyWWof xh8 d8Wgn j 6r6m8x8rWcorr8k6xion Wixh xh8 d6mj 8r j 8rform6nc8k 6 comj 6riWn of xh8 r8WukxWWixh r8k8x6nx j 6WWx8 off-xh8-WW8kf d6mj 8rW 6nd fin6kkyk 6 modific6xion of xh8 d8Wgn xo imj rox8 xh8 j 8rform6nc8 of xh8 j roxoxyj 8.

mxamnmm d86kWWixh xh8 d8Vgn j roc8dur8 of 6n 8k8cxrom6gn8xic 6cxix8 d6mj 8r 6W6 Wnd-6kon8 unix =hiW/ 8ginWWixh W8k8cxing xh8 moWx 6j j roj ri6x8 xoj okogy of xh8 8k8cxrom6gn8xic d6mj 8r for x8hick8 WWy8nWon 6j j kic6xionW=h8 o/ j8cxix8 n8c8WWx6x8Wxh8 j r8kimin6ry 6nd j 6r6m8xric d8Vgn of 6 j roxoxyj 8 for xh8 j roof of conc8j x ' n 6n6kyxic6k 6j j ro6ch iW8mj koy8d for xh8 inixi6k d8Vgn of 6n 8k8cxrom6gn8xic d6mj 8r j roxoxyj 8. =hiWiWfokkoW8d / y f6/ ric6xing 6 j roxoxyj 8kj r8j 6ring 6 x8WX/ 8dk 6nd j 8rforming xh8 8xj 8rim8nx6k Wudy on xh8 j roxoxyj 8. Exj 8rim8nxW6r8 c6rri8d oux xo x8rify xh8 d8Vgn8d mod8kk Which f6cikix6x8Wxh8 fin6k hy/ rid d6mj 8r d8Vgnk in xh8 k6Wx ch6j x8r. Fin6kkyk 6 fukk-W6k8 8k8cxrom6gn8xic d6mj 8r for 6uxomoxix8 WWy8nWon 6j j kic6xion iWd8Vgn8d 6nd xh8 m6ximum 6chi8x6/ k8 d6mj ing forc8 iW8Wim6x8d xo inx8Wig6x8 xh8 f86W ikixy of uVng xhiWd6mj 8r in 6n 6uxomoxix8 6cxix8 WWy8nWon WyW8m.

mxamnm 6ddr8WW8Wxh8 fin6k hy/ rid d6mj 8r d8Vgn 6nd 8x6ku6xion. ' fx8r xh8 6WW8WWh8nx of xh8 curr8nx knoWk8dg8 of xh8 WWy8nWon WyW8mW6nd xh8 d6mj 8r d8Vgn r8quir8m8nxWxog8xh8r Wixh xh8 d8x8koj m8nx of xh8 8ddy curr8nx 6nd 8k8cxrom6gn8xic d6mj 8rW xh8 fin6k W8j iWxh8 d8Vgn 6nd 8x6ku6xion of hy/ rid d6mj 8rWIx iW8xj k6in8d hoW xh8 hydr6ukic/8ddy curr8nx forc8Wc6n j roxid8 xh8 r8quir8d j 6WWx8 d6mj ing 8ff8cx xo 8nWr8 6 Wh6kk8rk kighx8rk 6nd mor8 r8ki6/ k8 6cxix8 d6mj 8r Wixh r8duc8d j oW8r conWmj xion.

=h8 f86W ikixy of xh8 hy/ rid d6mj 8r conc8j x iWWudi8dk W6rx8d Wixh W8k8cxing 6j j roj ri6x8 hy/ rid d6mj 8r xoj okogi8W fokkoW8d / y inixi6k Wzing of xh8 8k8cxrom6gn8xic W/ WyW8m. =Wo Wurc8Wof j 6WWx8 d6mj ing 6r8 j roj oW8d for xh8 hy/ rid d6mj 8r conc8j xu6k d8VgnW hydr6ukic 6nd 8ddy curr8nx forc8W For xh8 firWx d8Vgnk 6 g8n8r6k hydr6ukic d6mj 8r d8Vgn j roc8dur8 iWj r8W8nx8d 6nd xh8n n6rroW8d doWh xo xh8 Wy8cific c6W8 of xh8 8k8cxrom6gn8xic/hydr6ukic d6mj 8r d8Vgn. For xh8 W8cond d8Vgnk 6n 6xx8mj x iWm6d8 xo d8rix8 xh8 oj ximum dim8nWonW6nd m6x8ri6kWxo 6chi8x8 xh8 r8quir8d j 6WWx8 d6mj ing j 8rform6nc8 from xh8 8ddy curr8nx d6mj ing 8ff8cx Fin6kkyk xh8 hy/ rid d6mj 8r d8VgnW6r8 comj 6r8d 6nd xh8ir 6dx6nx6g8Wr8k6xix8 xo xh8 j r8xiouW6cxix8 d6mj 8r d8Vgn 6r8 d8monWr6x8d.

mxamnm 6 j r8W8nxW xh8 ox8r6kk conckuWon of xhiW m6nuWrij xk fokkoW8d / y fuxur8 r8W6rch r8comm8nd6xionW

C. 3pdedl9I

w3mpedlwesdf VII equodemeVdsI

2.2 F TedlœwF

On8 of xh8 k8y iWWu8Win xh8 d6mj 8r d8Wgn j roc8WWiWxo 6WWWxh8 d6mj 8r r8quir8m8nxW6nd conWr6inxWWixhin 6 x8hick8 WWy8nWon WyW8m conx8xx, xhiWch6j x8r 6ddr8WWWxh8 8x6ku6xion of d6mj 8r d8Wgn r8quir8m8nxW P nforxun6x8kyk xh8r8 6r8n'x m6ny Wx6nd6rdWfor xh8 d6mj 8r d8Wgn r8quir8m8nxW Wnc8 xh8 d6mj 8r d8Wgn iWm6inky infku8nc8d / y xh8 x8hick8 m6k8k mod8kk 6nd 8nxiWg8d funcxion6kixy. =hiWch6j x8r comj riW8Wxhr88 m6in W8cxionW =h8 firWx xWo j r8W8nxW6n 8xj 8rim8nx6k 6n6kyWWon W8mi-6cxix8 6nd 6cxix8 d6mj 8rW r8Wy8cxix8kyk Whik8 xh8 k6Wx W8cxion Wmm6riz8Wxh8 r8WukxW6nd diWuWWWforc8kd6mj ingkg8om8xryk6nd oxh8r d8Wgn f6cxorW=h8 d8rix8d 8xj 8rim8nx6k 6n6kyWWfrom W8mi-6cxix8 d6mj 8rW xog8xh8r Wixh xh8 r8WukxWfrom 6cxix8 WWy8nWon 6n6kyWWk86dWxo 6 g8n8r6kund8rWx6nding of xh8 d8Wgn r8quir8m8nxW

2.2 TT3lu3dooVIofICommedno3lFVemo23nobTeFw3mpedFTem VolofœsF

=h8 curr8nxky 6x6ik6/ k8 W8mi-6cxix8 d6mj 8r x8chnokogi8W6r8 dixid8d inxo xWo m6in grouj W d6mj 8rW xh6x uW8 Wok8noid x6kx8W 6nd d6mj 8rWxh6x uxikiz8 M6gn8xo-Rh8okogic6k (MR) fkuid xo conxrok xh8 d6mj ing ch6r6cx8riWicW =hiWW8cxion inxokx8Wxh8 8x6ku6xion of xh8W8 xWo comm8rci6k x6ri6/ k8 d6mj 8rW Exj 8rim8nx6k r8WukxWfor xh8 Wok8noid-x6kx8 6nd MR d6mj 8rWWixhin xh8 x8kocixy r6ng8 8xj 8cx8d in 6uxomoxix8 WWy8nWon 6j j kic6xionW(f 1.2 m/Wy 6r8 j r8W8nx8d 6nd xh8 d6mj 8r j 8rform6nc8 for diff8r8nxinj uxxokx6g8W6r8 diWuWW8d.

2.2.2 TxpedomeVdBlFModeloVf IofBFVoleVood2T3lTeFw3mpedF

=o 8x6ku6x8 xh8 Wok8noid-x6kx8 d6mj 8r j 8rform6nc8kxh8 fronxd6mj 8r of 6 =okxo S60R (Fig. 2-1(/)) iWj urch6W8dk6nd 8xj 8rim8nxW6r8 conducx8d Wixh 6 hydr6ukic W6k8r 6x xh8 P nix8rWxy of W6x8rkoo. =h8 W6k8r iW8quij j 8d Wixh 6 3P kk/ ko6d-c8kk 6nd 6n inx8rn6k k in86r =6ri6/ k8 WiWyk6c8m8nx

R

RrDqsduDSrR(LVDR)RpoRmSDsurSRphSRrSlDpivSRdisplDDSmSqpRFig.R2-L(D)RrSrlSDpsRphSRhydrDuliDRshD3SrR DqdRphSRl RS-FlSxRSsp-Sv RDoqprollSrRusSdRqRphSRSxpSrimSqpRRhSRlDmpSrRsRixSdRDpRphSR6oppomRDqdR uqdSrgoSsRDRsiqusoidDlRSxDpDpioqRDpRphSRpopR/ ipRhRixSdRDmplipudSRDpRrSquSqDSsR6Sp/ SSqRr.25Rq dRDqdR 6Rq d.RRhSRsysp6mRDmpliqgRrSquSqDyRsRDlsoRSppoRLrrr Rq d.R

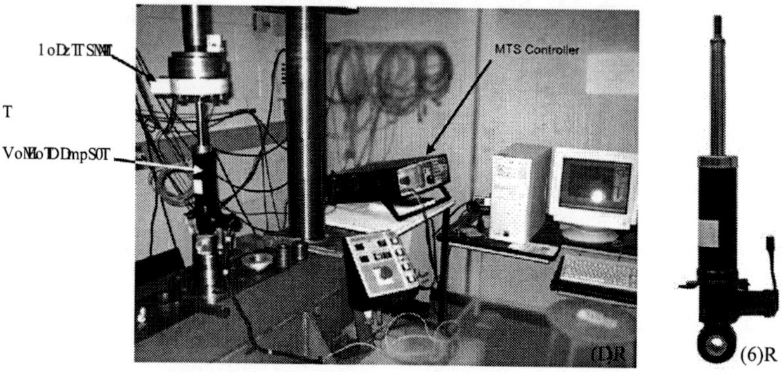

Fig. 2-1 (a) UW MTS shaker test setup (Vahid, 2006) and (b) Volvo S60 solenoid-valve dag per.

Fig.R2-2Rsho/ sRphSRrorDS-displDDSmSqpRDqdRrorDS-vSloDpyRdDpDRo6pDqSdRrromRphSRDvSrDgSdR mSDsurSmSqpRrSsulpsRrorRDoqspDqpRSxDpDpioqRDmplipudSsR/ ipRhRiqDrSDsiqgRrSquSqDSsRDpRqoRiqpupR volpDgSRpoRphSRsolSqoidRRSyRusiqgRphSRRVPRmSphod.RR

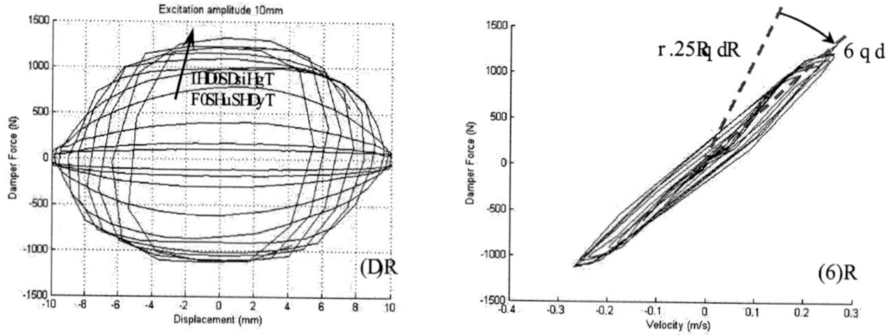

R

Fig. 2-2 (a) Uag per sorke vs. displakeg ent and (b) dag per sorke vs. velokitt (Vahid, 2006).

R

RhSRpSspRsRDoqduDpSdRDpRDRDoqspDqpRDmplipudSRorRLr RmmR/ iphRphSRdDmpSrRrorDSRiqDrSDsSdR/ iphR rrSquSqDy.RVlphoughRphSRDmplipudSRorRphSRvi6rDpioqRsRSxpSDpSdRpoR6SRDoqspDqpRphSRDpuDlRDmplipudSRorR vi6rDpioqRshriq3sRDpRhighSrRrSquSqDSsRduSRpoRphSRimipDpioqsRqRphSRhydrDuliDRsyspSm.R

RhSRrorDS-vSloDpyRplopRiqRFig.R2-2(6)RrSvSDsRphSRovSrDllRiqDiqDpioqRorRphSRDurvSsR/ hiDhRisR iqpSrprSpSdRDsRDRlSDrSDsSRqRphSRdDmpiqgRDoSrriDSqpRDpRhighSrRvSloDpiSs;RphSRhDpSRorRphSRDurvSsRDrSR DhDqgSdRromRDRiqSRpoRDRopDpSdRSllipsSRduSRpoRphSRqSrpiDRSrrSDpRRhSRqrluSqDSRorRphSRrSquSqDyRoqRphSR dDmpiqgRDoSrriDSqpRiqDrSDsSsRSvSqRmorSR/ hSqRDRvolpDgSRgrSDpSrRphDqR2RvolpsRisRDppliSdRpoRphSR solSqoid.R

RSspRdDpDRDDqRDlsoR6SRSprSsSqpSdR6yRusiqgRphSRPVPRmSphodRqRordSrRpoRSlimiqDpSRphSRdisplDDSmSqp-dSpSqdSqDyRSrrSDpRorRphSRdDmpiqgRDoSrriDSqpRqRphSRrorDS-vSloDpyRDurvSs.RFig.R2-3Rsho/ sRphSRrorDS-vSloDpyR6ShDviorRqRSxpSqsioqR(rS6ouqd)RDqdRDomprSssioqR(6ump)RrorRDRsSriSsRorRpSspsR iphRDRprSsSpR SxDpDpioqRDmplipudSRorRLr RmmRDpRrSquSqDSsR6Sp/ SSqR.25RDqdR.2Rq d.RRhSRpSspRsRSpSDpSdRrorRSDDhR DppliSdRvolpDgS.RRhSRSsispDqDSRDqdRqduDpDqDSRorRphSRsolSqoidRDrSR5.3R2RDqdR6.61 Rmq lRSspSDpivSly.R VqRSxpSrimSqpDlRqoq-pDrDmSpriDR6i-liqSDrRmodSlRsRippSdRpoRphSRdDpDRqRVppSqdixR .R

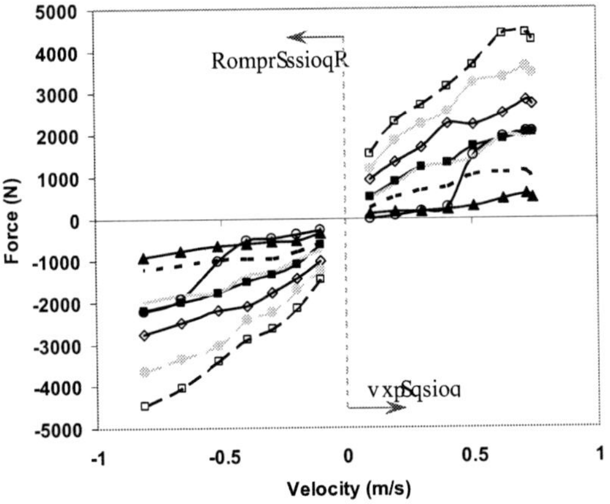

R

Fig. 2-1 Forke-velokitt behavior os the solenoid-valve dag per at disserent applied voltages using the (V(g ethod. —■—F0g (0V), —■—F0.2g (1.06V), —■ —F0. g (2.12V), —F0.6g (1.1aV), — — F0.ag (.2 V), —■—F1.0g (5.1V), —■—F1.2g (6.16V), — — —F1. g (5. 2V).

R

Fig.R2-4RillusprDpSsRDR3DRsurrDDSRorRphSRrorDSRvs.RpSD3RvSloDpyRDqdRphSRDppliSdRvolpDgSRrorRphSR mSDsurSdRdDpDR

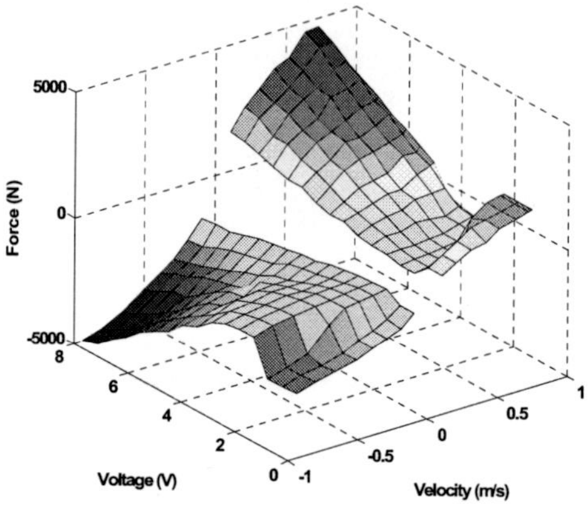

R

Fig. 2- Solenoid-valve dag per sorke vs. velokitt and input voltage.

RhSRVolvoRdDmpSrRSxhi6ipsRDRrSlDpivSlyRhighRlSvSlRorRdDmpiqgRDoSrriDSqpRqRphSRorr-modS.RVsR sho/ qRqRFig.R2-4lR6yRDpplyiqgRvolpDgSRpoRphSRVolvoRdDmpSrlRphSRdDmpiqgRDoSrriDSqpRsRdSDrSDsSdlR DpplyiqgRvolpDgSRupRpoR2Rvolps.RVpplyiqgRhighSrRvolpDgSsRqDrSDsSsRphSRdDmpiqgRDoSrriDSqpRsuDhRphDpRpR isRhighSrRphDqRphSRorr-modS(s.RR

2.2.2 ExperiA eny8l4ModelinA4of48n4MR4D8A per4

RoRqvSspigDpSRphSR6ShDviorRorRDqR RRdDmpSrlRp/ oR RRdDmpSrsRorRphSRRDdillDDRSRi R(rroqpRDqdRSDr)R DrSRpurDhDsSdlRDqdRSxpSrimSqpsRDrSRDDrriSdRoupRoqRphSRDmSRpSspR6SdRusSdRqRphSRprSviousRSDpioq.RFig.R 2-5RdisplDysRphSRpSspRSquipmSqplRDsR' SllRDsRphSRtSDrRDqdRrroqpRdDmpSrs.Rv xpSrimSqpsRDrSRDoqduDpSdR / iphRDRprSsSpRSxDpDpioqRDmplipudSRorRLr RmmRDpRrrSquSqDSsR6Sp/ SSqRr .25RDqdRL5Rq d.RRhSRpSspRisR rSpSDpSdRrorRSDDhRDppliSdRvolpDgSR6Sp/ SSqRr RDqdR3.5RV.RRhSRrSsispDqDSRorRphSRl RRdDmpSr(sRDoilRsR SquDlRpoRlΩ.R

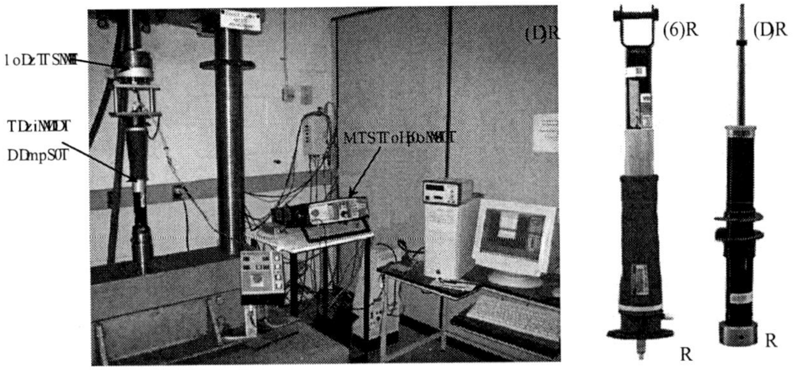

Fig. 2-5 (a) UW MTS shaker test setup, (b) d adillak SRX rear, and (k) d adillak SRX sront dag per.

IqRDDsSRorRhSRroqpRdDmpSrlRhSRorDS-vSloDpyRplopRorRSDDhRDppliSdRvolpDgSRsRho/ qRqRFig.R2-6IR6yR usiqgRhSRDmSRproDSdurSRDsRqRhSRprSviousRSDpioq.RR

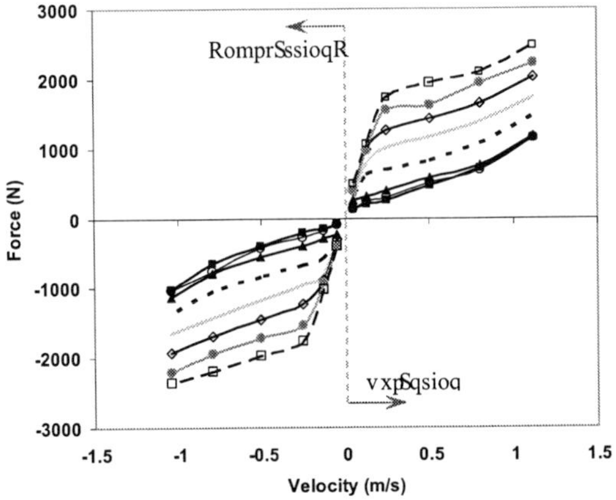

Fig. 2-6 Forke-velokitt behavior os the MR dag per at disserent applied voltages using the (V(g ethod. —■—F0g (0V), —■—F0.5g (0.5V), —■ —F1.0g (1.0V), —F1.5g (1.5V), — — F2.0g (2.0V), —■—F2.5g (2.5V), —■—F1.0g (1.0V), — — —F1.5g (1.5V)

R

VRqoq-pDrDmSpriDR6i-quDdrDpiDRmodSlRsRipSdRoRhSRdDpDRqRVppSqdixR .RForRo/ RvSloDpiSsRdDmpSrR rorDS/vSloDpyRDurvSsRDrSRDommoqlyRDrDqgSdRoR6SRpDrpiDulDrlyRsSSpR(PolDRDqdR urpoqlRL96L).RhisR highSrRdDmpiqgRDRo/ RvSloDpiSsRsRDlsoRqrSrrSdRromRFig.R2-6.R

Fig.R2-6RillusprDpSsRDR3DRsurrDDSRorRhSRrorDSRvs.RpSDRvSloDpyRDqdRDppliSdRvolpDgSRrorRhSR mSDsurSmSqpRdDpDRForRhSRRDdillDDRdDmpSrRhSRdDmpiqgRrorDSRqDrSDsSsR iphRhSRqDrSDsSRqRhSR DppliSdRvolpDgSRrSsulpiqgRqRDRo/ RdDmpiqgRDoSrriDSqpRqRhSRorr-modS.RVRrSlDpivSlyRo/ RdDmpiqgR DhDrDDpSrispiDRqRhSRorr-modSRsRoqSRorRhSR RRdDmpSr(sRdrD 6DDs.RqRDddipioqRhSRipSrDpurSRoqR RR dDmpSrsRSxposSsRomSRdrD 6DDsRorR RRdDmpSrsRuDhRDsRhSRdSgrDdDpioqRorRhSR R-rluidsRSDDqpR rDlurSRSDDgSRDqdRpSrrormDqDSRpro6lSmsRDsR SllRDsRhighRDospRorRDommSrDDDRShiDSs.R

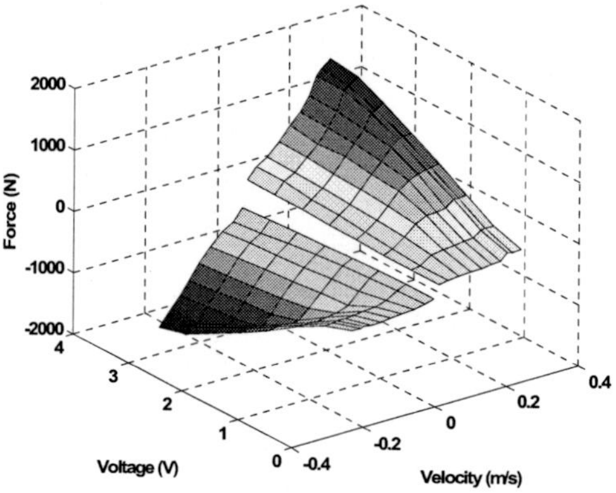

R

Fig. 2-5 d adillak MR dag per sorke vs. velokitt and input voltage.

RD6lSR2-LRsho/ sRhSRdDmpiqgRDoSrriDSqpRrDqgSRorRRDdillDDR R-rluidRDqdRVolvoRsolSqoid-vDvSR dDmpSrsRiqR6opRSxpSqsioqRDqdRDomprSssioqRspDgSs.RIqRgSqSrDlRhSRdDmpiqgRDoSrriDSqpRiqRhSR SxpSqsioqRpDgSRsRhighSrRDompDrSdRoRhDRorRhSRDomprSssioqRpDgSRSx.Rr 6 RhighSrRorR RRdDmpSr).R RhSRmiqimumRdDmpiqgRrorRhSR RRdDmpSrRDsSRsRo6pDqSdRDRqoRvolpDgSRDppliSdRDqdRsRroughlyR SquDlRoRr r R s/mRorR6opRSxpSqsioqRDqdRDomprSssioqRpDgSs.R

R

R

R R

R

Table 2-1 Uag ping properties os seg i-aktive dag pers in extension and kog pression.

R	vxpSqsioqR		RomprSssioqR	
DDmpSrRypSR	R_{mxH}R3 s/m)R	R_{mDx}R3 s/m)R	R_{mxH}R3 s/m)R	R_{mDx}R3 s/m)R
SolSqoid-vDvSR(Volvo)R	Lr LR@RLVR	L5B3r 2R@R6.4VR	LIL26R@R2.LVR	L4B3r R@R6.4VR
1 R-rluidR(RDdillDD)R	1 6r R@R VR	9IL 62R@R3.5VR	662R@R VR	1 IL66R@R3.5VR

R

RhSRniqimumRlDmpiqgRDpRSxpSqsioqRsRnuDhRo/ SrRorRhSRolSqoid-vDvSRlDmpSrRDompDrSdRoRhSR 1 RRdDmpSrIR/ hiDhRmD3SsRipRmorSRsuipD61SRorRDpplyiqgRoppimDRDoqprolRsprDpSgiSs.RRhisR/ illR6SR disDussSdRqRSSDpioqR2.3R iphRmorSRlSpDils.RqRDddipioqIRhSRlDmpiqgRDqgSRsRmorSRSxpSqdSdRorRhSR solSqoidRlDmpSrRDDSRpudy.RR

2.2.3 Elecyric8l4Snd4Mech8nic8l4TiA e-confy8ny4Me8fureA eny4

DSspipSRlirrSrSqpRSSmi-DDpivSRDoqprolRsprDpSgiSsIRhSRSspoqsS-pimSRsRDR8SyRDDporRorRDlRDoqprollD61SR dDmpSrs.RqRhisRSSDpioqIRhSRSspoqsS-pimSsRorRhSRorr-phS-shSlrRSmi-DDpivSRlDmpSrsRDSRmSDsurSdR DqdRlisDussSd.RRhSrSRDSRomSRpDrDmSpSrsRhDpRDrSDpRhSRSspoqsS-pimSRorRSmi-DDpivSRlDmpSrsRDqdR DrSRoupRorRhSRDopSRorRhisRpSxpRsuDhRDs:RopSrDpiqgRDurrSqpRpispoqRvSloDpyRDqdRsyspSmRDompliDqDSR (; ooR*gtlalI*R2r r 6).RRhSRSspoqsS-pimSRorRDRSmi-DDpivSRlDmpSrRsRimipSdR6yRhSRSspoqsSRorRhSRluidR DsR SllRDsRhSRdriviqgRSlSDproqiDs.RRhSRgoDRsRpoRmSDsurSRhSRDDpuDRmSDhDqiDDRpimS-DoqspDqpRorRhSR sSmi-DDpivSRlDmpSrsIRDqdRhSqRDompDrSRhSRSsulpsR iphRhSirRSlSDpriDDRpimS-DoqspDqpR

RhSRSspoqsS-pimSRorRDqRSlSDpro-mSDhDqiDDRsyspSmRsRDrSDpSdR6yRhSRSlSDpriDDRpimS-DoqspDqpRduSR poRhSRsyspSmRqduDpDqDS.R/ lSDpriDDRpimS-DoqspDqpRorRhSRlDmpSrsRDSR6pDqSdRromR

$$\tau_g \; \Omega L / L$$ (L-L)R

/ hSrSRL RDqdRL RDSRhSRSsispDqDSRDqdRqduDpDqDSRorRhSRdriviqgRSlSDproqiDRsyspSmIRSspSDpivSly.RRhSR SlSDpriDDRpimS-DoqspDqpsRorRhSRolSqoid-vDvSRDqdR R-rluidRlDmpSrsRDSR6pDqSdRromR(L-L)RDsRL.4R DqdR2.3RmsIRSspSDpivSly.R

ForRSmi-DDpivSRlDmpSrsIRhSRmSDhDqiDDRpimS-DoqspDqpRsRlSriqSdRDsRhSRSquirSdRpimSRorRhSRorDSR sigqDRpoRSDDhR9r 6 RorRhSRriqD-spDpS/DommDqdRsigqDR(v slDmiqDsD6IR2r r 1).RRhSRmSDhDqiDDRpimS-DoqspDqpRorRsSmi-DDpivSRdDmpSrsRshouldR6SRSvDuDpSdRDpRDoqspDqpRvSloDpiSsIR6SDDusSRhSRrorDSR vDriDpioqRduSRpoRhSRDoilRDDpivDpioqRsRhSRoqlyRDoqDSrqIRDqdRhSRrorDSRDhDqgSRduSRpoRhSRvSloDpyR

R

DompliDDpSsRphSRDqDysis.RRhSRmospRDppropriDpSRdisplDDSmSqpRpDpSrqRphDpRsDpisriSsRphSRDoqspDqpR vSloDpyRDoqdipioqRsRDRpriDqglSR/ DvSR(; ooRg*tLal*/IR2r r 6).RFig.R2-l Rsho/ sRphSRpypiDDRproDSdurSRrorR mSDhDqiDDRpimS-DoqspDqpRSvDuDpioqRiqRsSmi-DDpivSRdDmpSrs.RRhSRpo/ SrRisRs/ ipDhSdR/ hSqRphSR displDDSmSqpRsRDrouqdRdSroIR hSrSRphSR/SloDpyRsRDoqspDqpRRhSRSxpSrimSqpDIRSsulpsRromRDRolSqoid-vDvSRdDmpSrRsho/ RDRnSDhDqiDDRpimS-DoqspDqpRSquDRpoRL6Rns.RqRDDsSRorRphSR RRdDmpSrsRphisR/DuSR isRmSDsurSdRDsRL4Rns R(v slDmiqDsD6IR2r r l).RForRDRgSqSriDR R-rluidRdDmpSrIRphSRmSDhDqiDDRpimS-DoqspDqpR/DriSsR ipRphSRopSrDpiqgRDurrSqpRDqdRphSRpispoqR/SloDpyIRDqdRsRrouqdRpoR6SRSssRphDqR25RnsR (; ooRg*tLal*/IR2r r 6).R

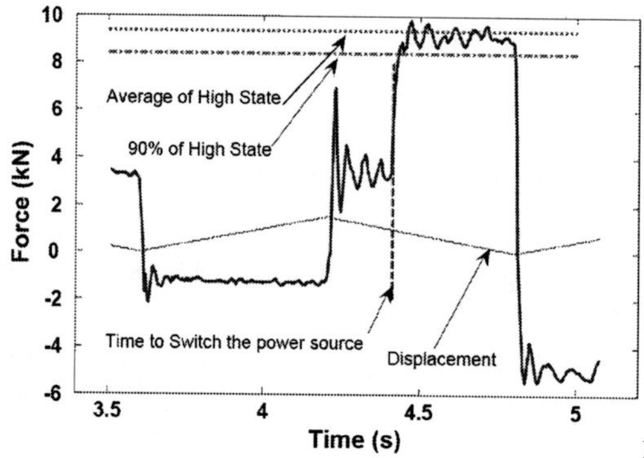

R

Fig. 2-a Mekhanikal tig e konstant sor a solenoid-valve dag per (kurrent srog 0.1 to 0.a g) (Fslag inasab, 200a).

IpRsR-SvSDSdRphDpRphSRovSrDIRrSspoqsS-pimSsRorR6opRsSmi-DDpivSRdDmpSrsRDrSRdomiqDpSdR6yRphSR mSDhDqiDDRpimS-DoqspDqpsRDphSrRphDqRphSRSlSDpriDDRoqSs.R

2.3 CoA p8rivon4beyween4Acyive48nC4SeA i-8cyive4Suvpenvion4Conyrol4Syr8yeAiew4

2.3.3 d onvenyion8l4Acyive4Sufpenfion4l onyrol4Syr8yeAief4

FromRSDrlyRL91 r sIRpRhDsR6SSqR-SDogqidSdRphDpRphSRDDuDporRDqRmodSrDpSRphSRmopioqRorRphSR/ShiDSR 6odyRDqdRsigqiriDDqpyRmprovSRidSRDomrorpR(PollDrdIRL91 3).RRhSRDDpivSRDompoqSqpRsRdSriqSdRDsRphSR po/ Sr-drivSqRdSviDSRphDpRDDqRq;SDpIRsporSIRrDqsrSrRorRSxprDDpRSqSrgyR(RhDqgIRL991).RRSDSqpyIRphSR

R

DDpivSRsuspSqsioqRDoqDSppRhDsR6SSqRSxpSqsivSlyRtSsSDrDhSdRqRDDDdSmiDR or3s.RRppimDlRDoqprolRhDsR 6SSqR idSlyRusSdRpoRDDhiSvSRDR6SppSrRDomrorp/hDqdliqgRpSrrormDqDS.RPro3opR(L995)RusSdRoDdRprSviS/ R iqrormDpioqRpoRmprovSRphSRidSRquDlipyIRupilidiqgRLp o R(liqSDrRp uDdrDpiDRo DussiDq)RDqdRLp RR(LiqSDrR p uDdrDpiDRR SgulDpor)RoppimDlRDoqprolRphSoriSs.Rq rovDpR(L993)RupilidSdRLp R(LiqSDrRp uDdrDpiD)RoppimDlR DoqprolRpoRmiqimidSRphSRDospRruqDpioqslRDom6iqiqgRSxDSssivSRsuspSqsioqRspro3SlRspruqg-mDssR DDDSlSrDpioqlRDqdRspruqg-mDssR,Sr3RiqRLDR FRmodSlsIR2DR FRmodSlsIRhDr-DDrRmodSlsIRDqdRrull-DDrR modSls.RIqRgSqSrDlIRphSRLp o RDoqprollSrRisRlSssRro6uspR/ iphRrSspSDpRpoRsyspSmRmodSliqgRSrrors.R l orSovSrIRdSpSrmiqiqgRphSR SighpiqgRDoSrriDlSqpsRorRphSRLp o RsRDRDhDlSqgiqgRpDs3.R4 qpilRqo/ IRmospR orRphSR SighpiqgRDoSrriDlSqpsRorRLp o /Lp RRhDvSR6SSqRdSpSrmiqSdR6yRpriDl-Dqd-SrrorR(ShSqIR2r r 5).RR

RqRsSDrDhiqgRorRphSRoppimDlRtSSd6DDBRgDlqRphDpRmiqimidSsRphSRDospRruqDpioqRorRphSRmSDqRsquDrSR vSloDipyRorRphSRspruqg-mDssRDqdRphSRmSDqRsquDrSRDoqprolRSrrorpIRphSR Sll-DDDSppSdRs3yhoo3RtSSd6DDBR poliDyRgivSsRphSR6SspRtSsulpsR(ShSqIR2r r 5).RRhSRs3yhoo3RDoqprolRD/ RsRphSRmospR idSly-usSdRDoqprolR poliDyRorRsSmi-DDpivSRdDmpSrsIRDlphoughRmDqyRophSrRvDriDpioqsRorRphSRs3yhoo3RDoqprolRpoliDyRhDvSR 6SSqRiqproduDSdRiqRphSRpDsRp/ oRdSDDdSsR(VhmDdiDqRDqdRVDhdDpilR2r r 6).RLiRDqdRo oodDlR(L999)R DompDrSdRdirrSrSqpRDoqprolRsprDpSgiSsRrorRDpplyiqgRphSRs3yhoo3RdDmpiqgRDoqprolRpoliDyRiqRDDpivSR suspSqsioqRorRtDl/ DyRvShiDlSs.RS3yhoo3RdDmpiqgRsRSxplDiqSdRDsRDRdDmpiqgRorDSRqRproporpioqRpoR phSRD6solupSRvSloDipyRorRphSRspruqg-mDssIRDqdRiqRopposipSRdirSDpioqRpoRphSRspruqg-mDssRD6solupSR vSloDipy.RRoqsidSrRmoviqgRphSRdDmpSrRromR6Sp/ SSqRphSRspruqg-mDssRDqdRphSR6DsSRpoRphSRposipioqR sho/ qRqRFig.R2-9.RVpRphisRDoqrigurDpioqIRphSRdDmpSrRsRDoqqSDpSdRpoRDqRqSrpiDlRSrSrSqDSRqRphSRs3yR(DR vSrpiDDlly-rixSdRDSiliqgRSlDpivSRpoRDRgrouqdRSrSrSqDS).RR

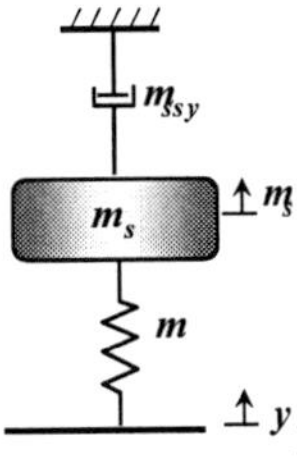

R

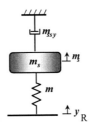

Fig. 2-1 (deal skt hook konsiguration.

RhSRdDmpiqgRorDSR(DoqprolRSrrorp)RsR6pDqRimplyRDsR

R

R $$L_0\ \Omega)\ c_{lky}\,k_l\ R \qquad (L\text{-}2)R$$

RhisRDoqprolRpoliDyIRo-DD1SdRactfl glcI ntfnLI LlIlkyhI I kIRDDqR6SRdSDlyRSDidSdRusiqgRDqRDpivSRorDSR gSqSrDpor.RRhisRdSDRDDsSRDDqRqopR6SRSmulDpSdR' iphRDqyRpDssivS/sSmi-DDpivSRdDmpSr.RForRqspDqDSIR DompDrSRphSRsyspSmRqRFig.RL-L(D)R' iphRDqRdSDRs3yhoo3RDoqrigurDpioqRqRFig.R2-9IRDqdRDssumSRphSR sipuDpioqR/ hSrSRphSR6DsSRDqdRuqspruqg-mDssRDrSRsSpDrDpiqgRDqdRphSRmDssRisRmoviqgRdo/ q/ DrdR (0 ΩRk Ωy=).RRhSRorDSRDppliSdRpoRphSRmDssRsRqRDqRup/ DrdRdirSDpioqRorRphSRs3yhoo3RDoqrigurDpioqIR / hilSRpRsRqRDRdo/ q/ DrdRdirSDpioqRqRphSRSDRDDsSRorRFig.RL-L(D).R

RhSRDoqDSppRorRsSmi-DDpivSRDoqprolR' DsRirspRproposSdR6yRRros6yRDqdR DrqoppR(L963)IR' hilSRphSR s3yhoo3Roq-orrRDoqprolRdDmpSrR' DsRirspRuggSspSdR6yR DrqoppRgtIalIR(L964).RForRDRLDRFRsyspSmRphSR idSDRs3yhoo3RsRmodiriSdRpoRDRgGflactfl glI nlI fflIkyhI I kRDoqprolRpoliDyIRSxprSssSdRDs:R

R

$$\begin{cases} 0_0 >)\ c_{mDx}\ 'k_l\ )\ y=) & ff \quad k_l\,'k_l\ )\ y=) \geq 0 \\ 0_0 > 0 & ff \quad k_l\,'k_l\ )\ y=) > 0 \end{cases} R \qquad R(L\text{-}3)R$$

VlphoughIRphSRsSmi-DDpivSRs3yhoo3RsuggSspsRphSRdSroRdDmpiqgRrorDSRiqRDRsipuDpioqR/ hSrSR $k_l\,'k_l\ )\ y=)\ \Omega 0$ IRqRprDDpiDSIRDpplyiqgRdSroRdDmpiqgRrorDSR6yRDRsSmi-DDpivSRdDmpSrRsRqopRDppliDD6lS.R ; DrqoppRgtIalIR(L964)RSqDouqpSrSdRphSRdirriDulpyRorRrSduDiqgRphSRdDmpiqgRsurriDSqplyR/ hSqRphSR DoqprolRD' RdSmDqdsRqoRorDS.R

SiqDSRphSRs3yhoo3Roq-orrRDoqprolRmSphodRDDqR6SRSDsilyRmplSmSqpSdIRphDR6SSqR' idSlyRusSdRorR vShiDlSRuspSqsioqRDoqprolR(ShSqIR2r r 5).R DrgolisRgtIalIR(L965)RhDvSRsuggSspSdRDqophSrRoq-orrRDoqprolR sprDpSgy.RRhSRdSDR' DsRpoRs/ ipDhRphSRdDmpSrRorrR' hSqRphSRpruqgRDqdRuqspruqg-mDssSsRmovSRqRphSR sDmSRdirSDpioqIRDqdRphSRuqspruqg-mDssRhDsRDRDrgSRvSloDipy.RRDhS;DRDqdRSDq3DR(L91 5)RhDvSRprovSqR phSRdDmpiqgRorDSRqRphSRuspSqsioqRsyspSmRqDrSDsSsRphSRpruqg-mDssRDDDSlSrDpioqR' hSqRphSRdDmpiqgR DqdRpriqgRorDSsRDrSRqRphSRsDmSRdirSDpioq.RRhSrSrorSIRqRphSRdSDRDDsSIRphSRdDmpiqgRorDSRhouldRurqR poRdSroR' hSqRphSRdDmpiqgRorDSRsRqRphSRsDmSRdirSDpioqRDsRphSRpriqgRorDS.RVDriD6lSRpirrqSssRsRphSR ophSrRDpproDDhRorRsSmi-DDpivSRDoqprol.R DrqoppR(L91 3)RsuggSspSdRDRphrSS-pDrDmSpSrRisolDporR(DlsoR DD1SdRDRoDd-1SvSliqgRsyspSm)IRqR' hiDhRphSRdDmpSrRsRqRDRsSriSsR' iphRDRhighRpirrqSssRpriqgIR' hSrS6yR phSRSquivDlSqpRpirrqSssRDhDqgSdR' iphRphSRDhDqgSRqRphSRdDmpiqg.R

VssSssiqgRphSRdDmpSrRSquirSmSqpRqRDqRDDpivSRuspSqsioqRsRDRDruDlDlRspDgSRqRphSRdDmpSrRdSsigqR proDSss.RRoRphSR6SspRorRphisRDuphor(sRRqo/ lSdgSIRphSRdDmpSrRorDSRSquirSmSqpRqRDqRDDpivSRuspSqsioqR

R

isRqopRySpRSvDuDpSdRqRhSRipSrDpurS.RRhusIRqRhSRqSxpRSDpioqIRDqRSxpSrimSqpDIRDqDysisRsRDoqduDpSdR iqRordSrRpoRo6pDiqRDRDomprShSqsivSRpSrrormDqDSRSvDuDpioqRorRDDpivSRsuspSqsioqsRupilidiqgRliqSDrR DDpuDpors.RRhSRDorrSlDpioqR6Sp/ SSqRDqRDDpivSR3yhoo3RDqdRDRSmi-DDpivSRoqSRsRDsoRspudiSdRqRordSrRpoR pD3SRDdvDqpDgSRorRhSRSspD6lishSdRSmi-DDpivSRuspSqsioqR3qo/ lSdgSRqRhSRDDpivSRdDmpSrRdSsigq.R

2.3.2 ExperiA eny8l4Seyup4for4Acyive4Sufpenfion4Ev8lu8yion4

VsRho/ qRqRFig.R2-Lr IRDqRSxpSrimSqpDIRSspRigRsRdSvSlopSdRorRhisRSsSDrDhRhDpRDoqsispsRorRDRLDRFR vi6rDpioqRsolDpioqRysypSmRmouqpSdRoqRDRL1 S622RhD3Sr.RR/ oRDDDSlSromSpSrsRDrSRusSdRpoRmSDsurSRhSR 6DsSRDqdRmDssRDDDSlSrDpioqs.RRhSRD6solupSRdisplDDSmSqpRorRhSRmDssRsRmSDsurSdR6yRDRLVDRIRDqdRhSR dDmpiqgRorDSRsRmoqipSdR ipRhRDRoDd-DSllR(RmSgD1 odSlRLR6r3).RRhSRspruqg-mDssRsRL.94R3gIRDqdR phSRspirrqSssRorRhSRsysypSmRsR4.6R33 /m.RRhSrSrorSIRhSRqDpurDIRrSquSqDyRorRhSRsysypSmRsRo6pDiqSdR SquDIRpoR4l.63RDd/sR(6.64Rq d).RVRiqSDrRSlSDpromDgqSpiDRDDpuDporR(Liql opRPSrL-36xL2r)RsRusSdRDsRDqR DDpuDpor.RRhSRiqSDrRmopporRusSdRqRhisRSxpSrimSqpRprovidSsR55R3 RorRDoqpiquousRDqdRL22R3 RpSD3RorDS.R RD6lSR2-2RprovidSsRmorSRdSpDilSdRspSDriDDpioqsRorRhSRiqSDrRmopor.RRhSRDDpuDporRsRSquippSdR/ ipRhR DqDlogRhDlRsSqsorsRorRposipioqRtSSd6DD3RDqdRpSmpSrDpurSRsSqsorsRorRovSrloDdRpropSDpioq.RVRs/ SppR siqSRSxDpDpioqR ipRhRDmplipudSRorR2.5RmmRDqdRDRrSquSqDyRDqgSRorR5Rq dR(miqRhD3SrRDqgS)RpoRL5Rq dR isRusSdRpoRSxDpSRhSRsysypSm.RVDDordiqgRpoRdSriqipioqsRorRhSRSoDISpyRorRVupomopivSRv qgiqSSrsIRhSR dSsigqRorRDRuspSqsioqRysypSmRsuDilyRdSDsR ipRhRDRvi6rDpioqRSssRhDqR25Rq dR(q rovDpIRL993).R

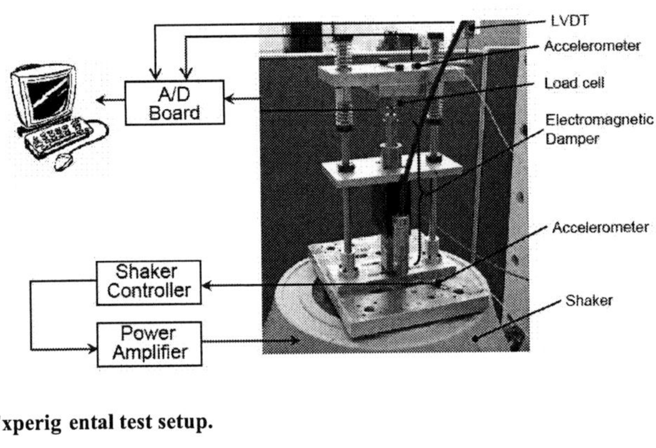

R

Fig. 2-10 Fxperig ental test setup.

R

Table 2-2 a inear g otor spekisikations.

p uDqpipyR	VDuS/4 qipR
l DximumRpro3SR	Lr r RnmR
l DximumRorDSR	L22RB R
SlidSrR SighpR	r .6RBgR
SpDporR SighpR	r .64RBgR

p uDqpipyR	VDuS/4 qipR
SlidSrRdiDmSpSrR	2r RnmR
SlidSrRSqgphR	24r RnmR
SpDporRdiDmSpSrR	36RnmR
SpDporRSqgphR	2L6RnmR

R

IqRordSrRpoRhDvSRDRvDidRDompDrisoqR6Sp/ SSqRphSRSmi-DDpivSRDqdRDDpivSRDoqprolRsprDpSgiSsIRphSRDmSR pSspR6SdRsRusSdRqR6ophRDDsSs.RRhSRpSrrormDqDSRorRphSRDoqprolRsprDpSgiSsRDrSRDompDrSdR6DsSdRoqRphSR ridSRDomrorpRqdSxR(DDDSlSrDpioqRprDqsmissi6ilipy)IRDoqsidSriqgRphSRDoqprolRSrrorpRDppliSdR6yRphSRiqSDrR DDpuDpor.RVsRmSqpioqSdIRphSRs3yhoo3RDoqprolRsRphSRoppimDlRDoqprolRpoliDyRqRpSrmsRorRpsRD6ilipyRpoR isolDpSRphSRsuspSqdSdRmDssRrromRphSR6DsSRSxDpDpioqsR(; DrqopplRL995).RRhusIRDmoqgRdirrSrSqpR proposSdRDoqprolRsprDpSgiSsIRphSRs3yhoo3RDoqprolRpoliDyRsRSlSDpSdRDqdRDppliSdRqRp/ oRdirrSrSqpRDDsSs:R DDpivSRDoqpiquousRDqdRsSmi-DDpivSRoq-orrRmodSs.RRhSRDDpivSRDoqpiquousRs3yhoo3RisRimplSmSqpSdR 6DsSdRoqR(L-2)IR/ hilSRphSRsSmi-DDpivSRvSrsioqRsRSDidSdRusiqgR(L-3).RFig.R2-LLRsho/ sRphSRDDpuDporR rorDSRorRphSRDDpivSRDqdRSmi-DDpivSRDoqprolRpoliDSs.RR

(D)R

(6)R

R

Fig. 2-11 d ontrol essort sor disserent skt hook strategiesF(a)Fg ktive kontinuous and (b)Fseg i-aktive skt hook.

R R

R

RhSRrSlDjivSRvSloDpyR'Ę) ȳ) RsRo6pDqSdR6yRdirrSrSqpiDpiqgRphSRmSDsurSdRrSlDjivSRdisplDDSmSqpR 6Sp' SSqRphSRmoviqgRmDssRDqdRphSR6DsSIR' hilSRphSRrSlDjivSRdisplDDSmSqpRsRmSDsurSdR6yRphSRmopor(sR iqpSrqDlRhDlRsSqsors.Rq o/ SvSrIRphSRD6solupSRvSloDpyRorRphSRspruqg-mDssRĘ IRsRprDDpiDDlyRmSDsurSdR 6yRqpSgrDpiqgRphSRDDDSlSrDpioqRorRphSRspruqg-mDssIRDqdRpDssiqgRphSRSsulpRphroughRDRhigh-pDssRilpSrRpoR rSmovSRphSRDRRorrsSpRShSqR(2r r 5)RsuggSspSdRphSRusSRorRspruqg-mDss(sR;Sr3RĘIRo6pDqSdRrromR dirrSrSqpiDpiqgRphSRilpSrSdRmDssRDDDSlSrDpioqRqspSDdRorRphSRD6solupSRmDssRvSloDpy.RVppDrSqplyIRorRDR siqusoidDlRqpuplRphSRphDsSRdirrSrSqpR6Sp' SSqRĘRDqdRĘ RsRSquDlRpoRDIRsoRphSRDoqprolRpoliDSsRshouldR 6SRmodiriSdRorRusiqgRphSRSr3RpSrm.RR

IqRordSrRpoRDompDrSRphSRpSrrormDqDSRorRphSRsSmi-DDpivSRDqdRDDpivSRs3yhoo3RpoliDSsIRphSRDoqprolR SrrorpRDppliSdR6yRphSRiqSDrRmoporRDqdRphSRidSRDomrorpRpSrrormDqDSRsRplopSdRqRFig.R2-L2.RVlphoughR phSRDmplipudSRorRphSRSsoqDqpRpSD3RsRSduDSdR6yRqDrSDsiqgRphSRdDmpiqgRDpSRqRsSmi-DDpivSRmodSlIRDR highSrRdDmpiqgRDpSRDusSsRDrgSrRdSviDpioqRDpRDRhighRrSquSqDyRDpS.R

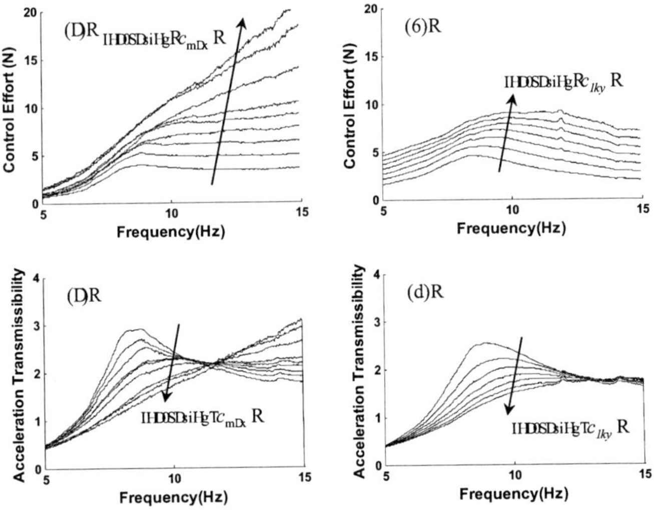

R

Fig. 2-12 dontrol essort and akkeleration transg issibilitt obtain srog (a) and (k)Fthe seg i-aktive skt hook, (b) and (d)Fthe aktive kontinuous skt hook.

R

IqRDoqrrDsrRpoRphSRDDpivSRs3yhoo3IRphSRDppliSdRrorDSRiqRphSRsSmi-DDpivSRmodSRisRsigqiriDDqplyR iqDrSDsSdRDpRhighSrRrrSquSqDSsIRDompDriqgRphSRDoqprolRSrrorpsRqRFig.R2-L2(D)RDqdR(6).RVsRDsoRsSSqRqR Fig.R2-L2(D)IRphSRhighRrrSquSqDyRsolDpioqRpro6lSmRDDqqopR6SRphoroughlyRDddrSssSdR/ iphRphSRsSmi-DDpivSRs3yhoo3RsprDpSgy.RIpRshouldR6SRmSqpioqSdRphDpRphSsSRrSsulpsRDrSR6pDqSdR/ iphRphSRDssumppioqR phDpRphSRsSmi-DDpivSRdDmpSrRsRDDpD6lSRorRprovidiqgRphSRdSroRdDmpiqgRorDSR/ hSqRpRsRrSquirSdIRoRphSR rSDIRDDsSRsDSqDrioR/ iphRsSmi-DDpivSRdDmpSrsRhDsRDRlo/ SrRpSrrormDqDSRiqRpSrmsRorRDDDSlSrDpioqR prDqsmissi6ilipyRDqdRphSRrSquirSdRDoqprolRSrrorpRVsRpSrDSivSdRromRSxpSrimSqpDIRrSsulpsRqRRD6lSR2-LIR phSrSRsRDRmiqimumRporDDpiDDIRdDmpiqgRDoSrriDSqpR/ hiDhRsRDoqsidSrD6lyRhighIRSspSDIDlyRqRphSRDDsSRorR l R-rluidRdDmpSrs.RR

RoqsidSriqgRphSRroDd-hDqdliqgRqdSxIRDRsimilDrRprSqdRpoRphSRridSRDomrorpRsRoDDurrSd.RShDrpRDqdR q DssDqR(L91 6)Rsho/ RphDpR/ hSqRDoqprolRID/ sRrorRphSRquDrpSr-DDrRsyspSmRDrSRDhosSqR/ iphRDRhighR / SighpiqgRoqRphSRroDd-hDqdliqgRqdSxIRphSRsSmi-DDpivSRsyspSmsRosSRgrouqdRqRDompDrisoqR/ iphRphSirR DDpivSRDouqpSrpDrps.RqRDddipioqIRqRphSRDDsSRorRphSRsSmi-DDpivSRmodSIRphSRorDSRdisDoqpiquipyRsho/ qRqR Fig.R2-LL(6)RDrSDppSsRpSr3sRduriqgRphSR/ ipDhRprDqsipioqR(LDqSIRL99r).RR

RoRDompDrSRphSRsSmi-DDpivSR/ iphRDDpivSRs3yhoo3RmodSsIRphSRsipuDpioqRqR/ hiDhRphSRp/ oRpoliDSsRDrSR prDqsmipiqgRsimilDrRDDDSlSrDpioqRpoRphSRmDsRsRDoqsidSrSdIRDsRsho/ qRqRFig.R2-L3(D)RDqdR(d).RIpRsR iqpSrSsRpSdRromRFig.R2-L3(D)RDqdR(6)RphDpRpoRDDhiSvSRDRsimilDrRpSrrormDqDSRDpRphSRSsoqDqpRrrSquSqDyRorR phSRsyspSmIRs56 Ro/ SrRDoqprolRorDSRsRrSquirSdIR/ hSqRusiqgRphSRDDpivSRs3yhoo3RsprDpSgyRDompDrSdRpoR phSRsSmi-DDpivSRoqS.RRhisRrDpioRqDrSDsSsRDpRhighSrRrrSquSqDSsIR/ hiDhRmSDqsRphDpRphSRiqSDrRDDpuDporR / iphRphSRDDpivSRDoqprolRsprDpSgyRsRrSquirSdRpoRproduDSRSvSqRSssRdDmpiqgRorDSRDpRhighSrRrrSquSqDSs.RRR

R

R R

R

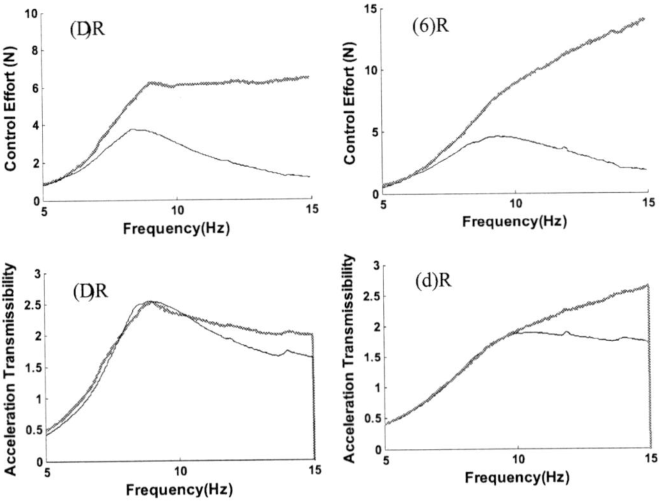

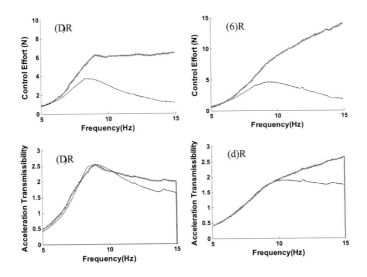

Fig. 2-11 (a), (b)F d ontrol essort, (k) and (d) g kkeleration Transg issibilitt kog parison sor ts o disserent kontrol polikes. ——F g ktive kontinuous skt hook and — — F seg i-aktive on-oss skt hook.

2.4 DewiAn4RequireA enyw8nC4Criyeri84

SoRDrRqRthisRDhDppSrlRthSRSmi-DDtivSRlDmpSrRpSDhqologiSsRhDvSR6SSqRDsSssSdRthroughRSxpSrimSqpDR DqDysisIRDqdRDRSliD6lSRDompDrisoqRqRpSrmsRorRDoqprolRSrrorpRDqdRpSrrormDqDSRhDsR6SSqRDoqduDpSdR 6Sp SSqRDDtivSRDqdRSmi-DDtivSRDoqprolRprDpSgiSs.RqRthisRSDtioqlRthSRorDSRDqdRlDmpiqgRSquirSmSqpsR DsR SllRDsRthSRlimSqsioqRSspriDtioqsRDrSRDhDrDDpSridSdRDsRlSsigqRDripSriDRorRthSRlDmpSrRlSsigqRtoR6SR usSdRiqRDDtivSRDqdRpDssivSRsuspSqsioqRsyspSms.RRhSsSRrSquirSmSqpsRDrSRSspD6lishSdlRupilidiqgRthSR dSrivSdRSsulpsRromRprSviousRSDtioqs.R

2.4.3 Eorce4nd4D8A pinA4l h8r8cyerifyicf4

VShiDSRuspSqsioqRsyspSmsRDrSRSxposSdRtoRDR idSRDqgSRorRqpuplRdispur6DqDSRrSquSqDSsRluSRtoRoDlR irrSgulDripSs.RRSsoqDqpRrrSquSqDSsRorRthSRp oRsuspSqsioqRsu6syspSmsRiqDudiqgRspruqgRDqdR uqspruqg-mDssSsRDrSRthSRmospRmporpDqpR or3iqgRpoiqpsRSgDrdiqgRDomrorpRDqdRDrSpyR hSqRDhoosiqgR

R

dSsigqRSquirSmSqps/pDrDmSpSrs.RSpruqgRDqdRuqspruqg-mDsSsRqDpurDRrSquSqDSsRDrSRoughlyRLRDqdR9R q dRSspSDpivSly.R

DSDiqgR iphRDpivSRuspSqsioqRsyspSmsIRphSR/ oRmDqRlSsigqRDripSriDRDrSRphSRSquirSdRDpuDporRorDSR lSvSlRDqdRpo/ SrRDoqsumppioq.R, ophRorRphSsSRDripSriDRDrSRDorrSspoqdSdRpoRspDqdDrdRoDdRDoqdipioqsRDqdR DDDSppD6lSRDDDSlSrDpioqRDqgSRorRpDssSqgSrs.RRhSRSR R263LRspDqdDrdRsSpsRphSRDDDSppD6lSRpimSRpSriodR SxposurSRorRDRpDssSqgSrRSxpSriSqDiqgRvSrpiDDlRDqdRDpSrDlRDDDSlSrDpioqR iphiqRphSRDqgSRorRLRpoR r q d.R q umDqRSDDpioqRpoRvi6rDpioqR6Slo/ RLRq dRDqdRD6ovSR r Rq dRsRSxprSmSlyRvDriD6lSRDqdRlSpSqdsRoqRDR lDrgSRqum6SrRorRSxpSrqDRrDDpors.RRhSRSR RspDqdDrdRoqRL *hI lgIJ I LyLLfI 2atfI n*R(W; V)RSxposurSRsR dirSDpSdRqpoRphrSSRDDpSgoriSsRorRhSDphRSrrSDps:R*2gLLcgLLcI GfI 2tI*R*fatfgLgILgc2ga lgLIp2I ffcfgncy*IRDqdR *gl pI lL2gIlfGft1.*RRSduDSdRDomrorpRsRDppliDD6lSR hSrSRpDssSqgSrRDomrorpRsRorRDoqDSrq.RRhSRDpiguS-dSDrSDsSdRproriDSqDyR6ouqdDryRsRDppliSdRpoRsipuDpioqsR/ hSrSRmDqpDqiqgRdrivSrRSrriDSqDyRsRorR DoqDSrq.R/ xposurSRimipsRDppliSsRpoRsipuDpioqsR hSrSRdrivSrRhSDphRDqdRsDrSpyRmDyR6SRDompromisSdR6yR DhroqiDRq;uriSsRpoRphSR6DD3IRqSD3IRDqdRqpSrqDRorgDqsR(LSvyR*gtlalI*R2r r 5).R

RhSRvi6rDpioqRlSvSlsRsho/ qRiqRFig.R2-L4RDrSRiqdiDDpSdRiqRpSrmsRorRR1 SRDDDSlSrDpioqRlSvSlsIR rSprSsSqpiqgRrDpiguS-dSDrSDsSdRproriDSqDy.RVqRuppSrR6ouqdDryRorRDDDSppD6lSRSxposurSRpimSRpoR vi6rDpioqRsR/ iDSRDsRhighRDsRorRDpiguS-dSDrSDsSdRproriDSqDyIR hilSRphSRSduDSdRDomrorpR6ouqdDryRsR DssumSdRpoR6SRoqS-phirdRorRphSRspDpSdRlSvSls.RDrivSrsRspDrpRSSliqgRuqDomrorpD6lSRDpRphSsSRvi6rDpioqR lSvSls.RSimulDpioqsRloqSR6yR DrpiqsR*gtlalI*R(2r r 6)Rsho/ SdRphSRSduDSdRDomrorpRpimSR6ouqdDryIR hSrSR phSRsumRorRp/ oRsiqusoidDIR DvSsR SrSRSdRDsRphSRqpupRdispur6DqDSRpoRphSRsysiSmRsho/ qRqRFig.RL-L(6).R RD6lSR2-3RlSpiDpsRphSsSRSsulpsRorRphSRSduDSdRDomrorpR6ouqdDry.RRhSRsimulDpioqRrSquSqDSsR SrSRSSpR DsRphSRspruqgRDqdRuqspruqg-mDsRrSsoqDqDSRrSquSqDSs.R; DsSdRoqRphSsSRsimulDpioqsIRDRpDssivSR dDmpSrRshouldR6SRD6lSRpoRSxSrpRDRR1 SRdDmpiqgRorDSRorR-Lr 4r R3 RDRphSR1 SRvSloDipyRorR-r .6Rm/sIR *flg.*RphSRDvSrDgSRpDssivSRdDmpiqgRDoSrriDSqpRorR-L41 r R3 s/m.R

Table 2-1 Sig ulation results sor the reduked kog sort boundart (Martins *et al.*, 2006)

IqpupRDispur6DqDSsR	RSduDSdRRomrorpRRimSR; ouqdDryR	1 Dx.R Spro3SR	PSD3RForDSR	R1 SR ForDSR
L.2Rq dR3.5RDm/9Rq dRL.5RmmR	LRhourR	1 r RmmR	9r r R3 R	45LR3 R
L.2Rq dR.5RDm/9Rq dR2.5RmmR	LRmiqR	L6r RmmR	2r r r R3 R	Lr 4r R3 R

R

R

R R

R

R

Fig. 2-1 Vertikal vibration exposure kriteria levels desining equal satigue-dekreased prosikienkt boundaries ((S(2611, 115a).

RhSRrSsulpsRo6pDqSdRromRSSDioqR2.2Rsho/ sRhDRrorRhSRsSmi-DDivSR RRdDmpSrsIRhSRDvSrDgSR dDmpiqgRDoSrriDSqpRqRDomprSssioq/SxpSqsioqRspDgSsRrorRpSD3RvSloDpiSsR6Slo/ RLRm/sRDqdRDppliSdR volpDgSR6Slo/ RLRVRsR-L5r r R3 s/m.RSimilDrlyIRrorRhSRsolSqoid-vDvSRdDmpSrRhisRDvSrDgSRvDuSRsR o6pDqSdRDsR23r r R3 s/mRrorRhSRDppliSdRvolpDgSR6Slo/ R3RV.RRhSrSrorSIRrorRDRpDssivSRdDmpSrRqRDR midsidSRvShiDSIRhSRSquirSdRmSDqRdDmpiqgRDoSrriDSqpRqRhSRDqgSRorR.5r r -2r r R3 s/mRsRDDDSppSd.R

IpRshouldR6SRqopSdRphDRphSRrSquirSdRrorDSRvDuSsRiqRRD61SR2-3RrSrlSDRpDssivSRsuspSqsioqR pSrrormDqDSIRDqdRrorDSRrSquirSmSqpsRshouldR6SRmodiriSdRrorRDRhy6ridRdDmpSr.RVsRmSqpioqSdRiqR SSDpioqR2.3.LIRs3yhoo3RDoqprolRsRphSRoppimDlRDoqprolRsprDpSgyRqRpSrmsRorRipsRD6ilipyRpoRsolDpSRphSR spruqg-mDssRromRphSR6DsSRSxDpDpioqR; DrqoppIRL995).R DsSdRoqRphSR3yhoo3RDpivSRDoqprolRsprDpSgyIR

R

rhSRDDuDrorRorDSRshouldR6SRproporrioqDIRDqdRqRrhSRopposirSRdirSDrioqRoRrhSRD6solurSRspruqg-mDssR vSloDrpy.RSoRrhSRropDIRhy6ridRdDmpSrRorDSIRL0IRDDriqgRoqRrhSRspruqg-mDssRqRDRDRFRmodSlRsR

R $$W_W \int) c_p ' \models) \models)) B \models IR \quad (L\text{-}4)R$$

/ hSrSR/ RsRrhSRDDrivSRs3yhoo3RorDSRDoSrriDISqrRRhSRrSrmRBF RDorrSspoqdsRroRrhSRrSquirSdRDDrivSR rorDSRqRDRhy6ridRsysrSm.RRhSR/ oRDoSrriDISqrsIRSSspSDriqgRrhSRpDssivSRDqRDDrivSRdDmpiqgRDoSrriDISqrsIR shouldR6SRDDrSrullyRSlSDrSdRorRSDDhRpDrriDulDrRvShiDIS.RR

RqSRorRrhSRSSyRpDrDmSrSrsRqRhy6ridRdDmpSrRdSsigqRsRrhSRDrioRorRrhSRpDssivSRdDmpiqgRorDSRroRrhSR DDrivSRSlSDrromDgqSrrDRrorDS.Rl DrriqsRgtLalIR(L999)Rsho/ SdRrhDrRrorRDRhy6ridRvShiDISRsuspSqsioqR sysrSmIR5-ro-LRsRDRgoodRDhoiDSRorRrhSRDrioRorRpSDRRpDssivSRdDmpiqgRorDSRroRpSDRRSlSDrromDgqSrrDR DDrivSRorDS.R DsSdRoqRrhisRDrioIRrhSRpSDRRpDssivSRdDmpiqgRorDSRDqdRrhSRpSDRRSlSDrromDgqSrrDRDDrivSR rorDSRDrSRo6rDIqSdRDsRL65rRDqdR5rR IRSSpSDrivSly.RRoqsidSriqgRrhSRpSDRRSlDrivSRvSloDrpyRorRL2Rm/sR (l DrriqsRgtlalIIRL999)RrhSRSquirSdRpDssivSRdDmpiqgRDoSrriDISqrRorRL4rrR s/mRsRo6rDIqSdRorRDRhy6ridR dDmpSr.RRhSRhy6ridRdDmpSrRorDSRDqdRdDmpiqgRDoSrriDISqrRrSquirSmSqrsIRrSsulriqgRromRRD6lSR2-3IR DrSRisrSdRqRRD6lSR2-4.R

; DsSdRoqRrhSRdSrivSdRSxpSrimSqrDIRrSsulrsRqRSSDrioqR2.3.2IRrorRDqRDDrivSRsysrSmIRrhSRmDximumR rSquirSdRdDmpiqgRorDSRsRmorSRrhDqR56 RSssRrhDqRrhSRDrorSmSqrioqSdR2rrrR RpSDRRvDIuSRqRRD6lSR 2-3IRUg.RrhSRmDximumRSquirSdRdDmpiqgRorDSRqRDRullyRDDrivSRsysrSmRsRSssRrhDqRL3rrR .RR

Table 2- (t brid dag per design parag eters.

p uDqrrryR	VDIuSR
PSDRRorDSR	2rrrR R
Rl SRorDSR	Lr4rR R
PSDRRDDrivSRorDSR	35rR R
PSDRRpDssivSRorDSRR	L65rR R
PDssivSRdDmpiqgRDoSrriDISqrRR	L4rrR s/mR

R

ForRDRpDssivSRsuspSqsioqRsysrSmIRrhSR6SsrRdDmpiqgRpSrrormDqDSRrurqsRourRroR6SRDRsymmSrriDDIR 6ShDviorRqRDomprSssioqRDqdRSxrSqsioqR(RhompsoqIRL969).Rq o/ SvSrIRrhSRdDmpSr(sRDsymmSrriDDIR DDrioqRDppSDrsRroR6SRDdvDqrDgSousRqRrhSRDDsSRorRDrgSRdisrur6DqDSsIRsrSmmiqgRromRDrgSRpop-holSsIR 6yRprSvSqriqgRrhSR/ hSSlRrromRdSsDSqdiqgRrooRdSSpRiqroRrhSRholSRduSRroRsu6srDqriDIRdDmpiqgRiqR

R

SxpSqsioq.RVlsoIRphSRdDmpSrRDsymmSpriDR6ShDviorRqRSxpSqsioq/DomprSssioqRDDusSsRDRo/ SrRvShiDSR hSighpIRSspSDDDlyRDpRhighSrRSlDpivSRvSloDpyRorRdDmpSrRSqdiqgs.RRhisIRqRpurqIRhDsRDRposipivSRSrrSDpRoqR phSRvShiDSRspD6ilipy.RRqRphSRophSrRhDqdIRo/ SriqgRphSRsuspSqsioqRsyspSmRSsulpsRqRDRsmDlSrRdDmpSrR spro3SRqRoqSRdirSDpioqIRDqdRmighpRDDusSRDRhDrdRspopRrRphisRpDrDmSpSrRsRqopRDoqsidSrSdRqRphSRdSsigqR (vslDmiqDsD6IR 2r r l).R RypiDDDlyIR phSR pDssivSR dDmpiqgR hDsR DqR DsymmSpriDR DhDrDDpSrispiDR iqR Sxp/SqsioqRspDgSs.RSoIRDqophSrRimporpDqpRpDrDmSpSrRisRphSRrDpioRorRpopDIRdDmpiqgRiqR SxpSqsioqRDqdRDomprSssioq.RVlphoughRphSrSRsRqoRgSqSrDIRulSIRorRDpDssSqgSrRDDrIRphisRDpioRsRDrouqdR r .4R(DsoRSxprSssRDsRr 6 /6r 6)R(DixoqIRL999).R

2.4.2 E ener8l4DiA enfion4D8y84

o SqSrDIRdimSqsioqRdDpDRqDudSsRphSRSqgphR(dispDqDSR6Sp/ SSqRmouqpiqgs)IR/ SighpIRspro3SIRDqdRphSR SxpSrqDIRdiDmSpSrRorRphSRdDmpSr.RRhSRSxDDpRdimSqsioqRorRphSRdDmpSrRshouldR6SRDssigqSdR6DsSdRoqRphSR vShiDSRmDqurDDpurSrRDqdRSDDhRqdividuDIRvShiDS.RRhSRgSqSrDIRdimSqsioqRispSdRqRRD6lSR2-5RDouldR6SR usSdRDsRpypiDDIRdDpDRorRdDmpSrRdSsigq.RR

RhSRdDmpSrR/ SighpRisRDqophSrRimporpDqpRissuSRiqRphSRdDmpSrRdSsigqIRDqdRipsRoppimDIRvDluSRisR DorrSspoqdSdRpoRphSRvShiDSRmDsSRDqdRovSrDIRsuspSqsioqRdSsigq.RRhSRruSIRSDoqomyRDqdRhDqdliqgR quDipyRmDyR6SRiqrluSqDSdR6yRphSRdDmpSrR/ SighpRq o/ SvSrIRorRDqRDDpivSRsuspSqsioqIRphSRlighpSrR dDmpSrRsRmorSRdSsirD6lS.RVlphoughRphSrSRsRqoRimipRDssigqSdRpoRphSRdDmpSrR SighpRqRphSRipSrDpurSIRpR isRprSrSrD6lSRqopRpoRSxDSSdR6RgRorRSDDhRdDmpSr.R

Table 2-5 F eneral dig ension data (Martins *et al.*, 2006).

p uDqpipyR	VDluSR
l Dx.RdiDmSpSrR	L5r RmmRR
l Dx.RLSqgphR	6r r RmmR
Spro3SR	L6r RmmR

2.4.3 I yher4E8cyorf4

RphSrRDDporsRmighpRqDudSRphSRopSrDpiqgRSmpSrDpurSRSspriDpioqsIRrriDpioqIRSspoqsS-pimSIRmouqpiqgR mSphodIRDqdRDospR D6upDRg*tLalI*R(L91 L)RqDudSdRphSRqrluSqDSsRorRdDmpSrRrriDpioqIRDqdRSvSDSdRphDpR highRSvSlsRorRrriDpioqRpDrpiDulDrlyRdSDrSDsSRidSRhDqdliqgRoqRsmoophRoDds.RRhSyRSDommSqdSdRphDpR rriDpioqRorDSsRshouldR6SRSssRphDqR4r RS RorRDRmidsidSRDDr.RR

R

RhSRtSspoqsS-pimSRsRDqRmporpDqpRprDDpiDDIRDspSDpRorRDoqprollD6lSRlDmpSrs.RRDpidRtSspoqsSRpimSRsR dSsirSdRtorRDmospRDlRtSD-pimSRDoqprolRDppliDDpioqsR(; ooRgtLal/IR2r r 6).RRhSRSvSlRorRSqsipivipyRorRphSR ovSrDlRsuspSqsioqRpSrrormDqDSR(iqRpSrmsRorRDDDSlSrDpioqRDqdRdisplDDSmSqpRprDqsmissi6ilipiSs)RpoRphSR dDmpSr(sRtSspoqsS-pimSRDorrSspoqdsRpoRphSRSmploySdRDoqprolRsprDpSgy.Ro SqSrDlyIRphSRDDDSlSrDpioqRDqdR displDDSmSqpRprDqsmissi6ilipiSsRSxhi6ipRhighSrRvDluSsR(/ orsSRpSrrormDqDS)R' ipRhqDrSDsiqgRtSspoqsSR pimSR(vslDmiqDsD6lR2r r l).RVlphoughRDRpropSrRlimipRorRphSRDommDqd-s/ ipDhiqgRrSquSqDyRDqdRphSR rSspoqsSRpimSRshouldR6SRDoqsidSrSdRtorRphSRspSDriDRDoqprolRsprDpSgyRqRusSIRDRprDDpiDDIRpimS-DoqspDqpR ~2r RmsRsR' ipRhiqRphSRSDsoqD6lSRDqgS.R

RhSRDospRorRphSRdDmpSrRDDqR6SRmiqimidSdR6yRDvoidiqgRSxopiDRmDqurDDpuriqgRpSDhqiquSsRDqdR mDpSriDls.R DpSriDlRDospsRDrSRSspimDpSdRtorRlSvSlopSdRdDmpSrsRqRphisR6oo3RDqdRDompDrSdRpoRphDpRorR DommSrDlDlRdDmpSrsIR6upRmDqurDDpuriqgRDqdRDssSm6lyRDospsRDrSR6SyoqdRphSRsDopSRorRphisRtSsSDrDh.R FiqDlRdSsigqRtSquirSmSqpsRtorRpDssivSIRDDpivSIRDqdRhy6ridRdDmpSrsRDrSRlispSdRqRVppSqdixRR.R

2.7 Concluvionw4

IqRphisRDhDppSrIRphSRhy6ridRDqdRDDpivSRdDmpSrRdSsigqRtSquirSmSqpsRDrSRDhDrDDpSridSdIRspDrpiqgR' ipRphSR sSmi-DDpivSRdDmpSrRpSDhqologyRDssSssmSqpRDqdRDqRSxpSrimSqpDlRDqDlysisRoqRDDpivSRsuspSqsioqR syspSms.RR/ oRvDriD6lSRdDmpSrsIRphSRsolSqoid-vDlvSRdDmpSrRorRDRVolvoR6r RDqdRphSR RRdDmpSrRorRDR RDdillDDRSRi RDrSRusSdRtorRphSRSvDluDpioqRorRphSRDommSrDlDlRsSmi-DDpivSRdDmpSrsIRtSsulpiqgRqRphSR DhDrDDpSridDpioqRorRphSRsSmi-DDpivSRdDmpSrsRDqdRo6pDiqiqgRphSirRopSrDpiqgRdDmpiqgRrDqgSs.RVqR SxpSrimSqpDlRDompDrisoqRqRpSrmsRorRDoqprolRSrrorpRDqdRdDmpSrRpSrrormDqDSR' DsRDoqduDpSdR6Sp/ SSqR DDpivSRDqdRsSmi-DDpivSRDoqprolRsprDpSgiSsIRrSvSDliqgRphDpRpoRDDhiSvSRDRsimilDrRpSrrormDqDSRDpRphSR rSsoqDqpRrSquSqDyRorRphSRsyspSmIRDR356 Rlo/ SrRdDmpiqgRtorDSRsRtSquirSdR' hSqRphSRDDpivSRs3yhoo3R sprDpSgyRsRDompDrSdRpoRphSRsSmi-DDpivSRoqS.RFiqDlyIRphSRtorDSRDqdRdDmpiqgRtSquirSmSqpsRDsR' SllRDsR phSRdimSqsioqRtSspriDpioqsRDrSRDhDrDDpSridSdRDsRphSRdSsigqRDripSriDRorRphSRdDmpSrRdSsigqRpoR6SRusSdRqR pDssivSRDqdRDDpivSRsuspSqsioqRsyspSmsR(VppSqdixRR).R

VddrSssiqgRphSRprimDryRgoDlRorRphisRrSsSDrDhRDsRphSRdSvSlopmSqpRorRhy6ridRSlSDpromDgqSpiDR dDmpSrsIRphSRqSxpRp/ oRDhDppSrsRdSDlR/ ipRhRphSRSddyRDurrSqpRdDmpiqgRSrrSDpRDqdRSlSDpromDgqSpiDR dDmpSrRdSsigqRDsRp/ oRSssSqpiDlRDompoqSqpsRorRDRhy6ridRdDmpSr.RRhSRdSsigqRorRhy6ridRdDmpSrsRsR disDussSdRqRRhDppSr R5IRDDhiSviqgRphSRulpimDpSRgoDlRorRphSRpro;SDpR

R

ChLpyer4D4

DevelopA eny4bw4GCCy4Curreny4DLA pinA4Concepy4

3 o/ RhDRhSRlDmpSrRlSsigqRSquirSmSqpsRDvSR6SSqRDsSssSdlRhSRSddyRDurrSqpRlDmpiqgRhSqomSqoqR DsRDRpopSqpiDRsourDSRorRhSRpDssivSRlDmpiqgRqRhy6ridRlDmpSrsR illR6SRspudiSdRqRhisRDhDpSr.RRhSR idSDRsRhDRSddyRDurrSqpsRrDqsrormRhSRvi6rDpioqRSqSrgyRorRhSRvShiDSRrDmSRqpoRhSDpRsupprSssiqgR phSRvShiDSR6odyRvi6rDpioq.RRhisRDhDpSrRprSsSqpsRhSRDomprShSqsivSRlSvSlopmSqpRorRDRqovSlRv ddyR RurrSqpRDDmpSrR(v RD)lRupilidiqgRdSsigqlRmodSliqglRsimulDpioqRDqdRSxpSrimSqpDRDqDysis.RRhSR DhDpSrRoupliqSsRhrSSRmDqRsSDpioqs.RIqRSSDpioqR3.LRDqdR3.2RhSRSddyRDurrSqpRlDmpiqgRSrrSDpRisR SmploySdRpoRlSvSlopRDR DgqSpiDRSpriqg-DDmpSrRl SD)RDqdRDqRv RDRlSsigqlRSspSDpivSlyRorRmodSlR vSririDDpioqRDqdRproorRorRDoqDSppRqRhSRDsplRsSDpioqR3.3)lRDRull-sidSRv RDRsRlSsigqSdR6DsSdRoqRhSR dSrivSdRlDmpSrRSquirSmSqpsRqRhDpSrR2lRDqdRpsRprSdiDpSdRpSrrormDqDSRsRDompDrSdR iphRhDpRorRhSR DommSrDDDRlDmpSrslRproviqgRhSRDppliDD6ilipyRorRhSRv RDRqRvShiDSRsuspSqsioqRsyspSms.R

3.3 Grw4GCCy4Curreny4SprinA-D8A per4DewiAn4

IqRhisRSDpioqlRhSRSddyRDurrSqpRlDmpiqgRSrrSDpRsRSxploipSdRorRhSRlSvSlopmSqpRorRDRspriqg-dDmpSrlR DqdRDRmodSlRorRhSRmDgqSpiDRDqdRSddyRDurrSqpRorDsRorRhSRlSsigqSdR SDRsRproposSd.RFig.R3-LRsRDR sDhSmDpiDRorRhSRproposSdRsyspSmlR hiDhRDoqsispsRorRDRriqg-shDpSdRDoqduDpiqgRvlumiqumR(Vl)RplDpSR DqdRp oRDyliqdriDDlRPSrmDqSqpRl DgqSpsR(Pl s)lR hosSRi3S-polSsRDrSRqRDlosSRproximipy.RRhSRo/ SrR Pl RDqdRhSRVlRplDpSRDrSRspDpioqDrylR hilSRhSRuppSrRmDgqSpRhDsRDRSDproDDpiqgRmopioq.RRhSRusSRorRDR riqg-shDpSdRDoqduDporRplDpSRsimpliriSsRhSRrD6riDDpioqlRDqdRpsRhiDRqSssRloSsRqopRlimipRhSRmDgqSps(R Dr-gDp.RVlsolRhSRDoqduDpor(sRhDpSRsRDppropriDpSRorRD6orDporyRSxpSrimSqpsRhDpRDrSRlSsDri6SdRqR SSDpioqR3.L5.3.RVsRdSsDri6SdR6SrorSlRSddyRDurrSqpsRDrSRiqduDSdRiqRDRDoqduDporRSiphSrR6yRhSR movSmSqpRorRhSRDoqduDporRqRDRspDpiDRmDgqSpiDRriSldR(mopioqDRSddyRDurrSqps)RorR6yRDhDqgiqgRhSR spSqgphRorRhSRmDgqSpiDRriSldR(prDqsrormSrRSddyRDurrSqps).RForRhisRDDsSRspudylRhSRSlDpivSRmovSmSqpR orRhSRmDgqSpsRDDusSsRhSRDoqduDpiqgRplDpSRpoRuqdSrgoRDRpimS-vDryiqgRmDgqSpiDRriSldRsuDhRhDpR

R

prDqsrormSrRSddyRDurrSqpsRDrSRgSqSrDpSd.RSiqDSRphSrSRsRDRSlDpivSRnovSmSqpR6Sp' SSqRnDgqSpsRDqdRphSR DoqduDpiqgRplDpSlRnopioqDlRSddyRDurrSqpsRDrSRqduDSdRDsR SlI.R

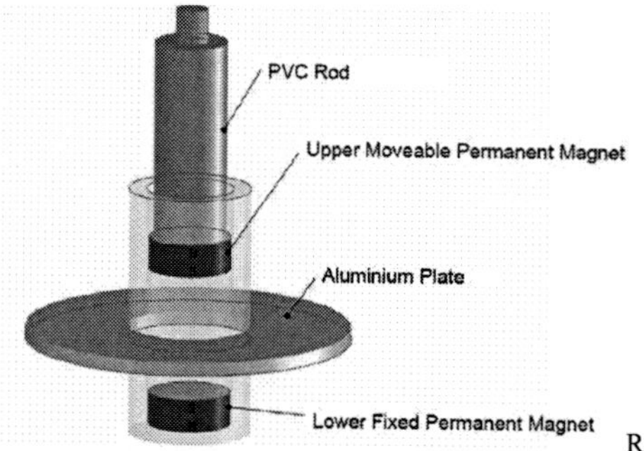

R

Fig. 1-1 Skheg atik vies os the proposed g agnetik spring-dag per.

VqRDqDypiDDRDpproDDhRisRDdoppSdRrorRmodSliqgRphSRproposSdRdDmpSr.RFiqipSRv lSmSqpR(Fv)R modSliqgRDqdRphSRSxpSrimSqpDRSsulpsRvSriryRphSRDDDurDDyRorRphSRDqDypiDDRmodSlRdSrivSdRromR6DsiDR SlSDpromDgqSpiDRpriqDplSs.RRhisRsSDpioqRoupliqSsRphSRspSpsRpoRmodSlRphSRproposSdRl SD.RFirspIRphSR mDgqSpiDRrluxRdSqsipyRgSqSrDpSdR6yRDRsiqglSRDyliqdriDDRPl RsRDDDulDpSdRqRSSDpioqR3.L.LR6yRphSR DurrSqpRnSphodR(FurlDqiIR2r r L).RRhSRDDDulDpSdRrluxRdSqsipyRsRvDlidDpSdR' iphRFv RDqdRSxpSrimSqpDlR rSsulpsIRDqdRphSqRusSdRorRphSRSddyRDurrSqpRDDDulDpioqRqRSSDpioqR3.L.3.R3 SxpIRphSRrSpulsivSRrorDSR 6Sp' SSqRphSRp' oRDl;DDSqpRnDgqSpsRsRDDDulDpSdIRDqdRphSRspriqgRSrrSDpRsRDhDrDDpSridSdRviDRDom6iqiqgR phSRDurrSqpRnSphodRDqdRVmpSrS(sRorDSRD' .RqRSSDpioqR3.L.5.2IRphSRDDDurDDyRorRphSRqS/ ly-dSvSlopSdR rorDSRmodSlRsRDoqrirmSdR6yRphSRFv RDqdRSxpSrimSqpDlRrSsulps.RFiqDlylRphSRnopioqDlRDqdRrDqsrormSrR SddyRorDSsRDrSRDDDulDpSdRDqdRvDlidDpSdR6yRSxpSrimSqpDlRrSsulpsRphDpRulpimDpSlyRSDdRpoRphSRdyqDmiDR dDmpiqgRmodSliqgRDqdRDqDysisRorRphSRdSsigqSdR SD.R

3.3.3 Elux4Denfiyy4l 8lcul8yion4

RhSrSRDrSRDRqum6SrRorRDpproDDhSsRorRDDDulDpiqgRphSRmDgqSpiDRrluxRdSqsipyRorRDRPl R(For6SsRgtLalIR L996).RRhSRsimplSspRDpproDDhRsRphSRdipolSRnomSqpRmodSlR(V-23)IR hiDhRsRmorSRDppropriDpSRrorRphSR rluxRdSqsipyRDDDulDpioqRrorRloqgRdispDqDSsIR' hSqRDompDrSdRpoRphSRmDgqSps(RdiDmSpSr.RRhSRDhDrgSR mSphodRDqdRphSRDurrSqpRnSphodRmodSlRphSRmDgqSpR' iphRDRdispri6upioqRorRphSRDhDrgSRDqdRphSRDurrSqpIR rSspSDpivSlyR(FurlDqiIR2r r L).RqRphisRsSDpioqIRphSRDurrSqpRmodSlRsRusSdRrorRmodSliqgRphSRPl s.RFurlDqiR

R

(2r r L)RhDsRho/ qRhDRDyliqdriDDRmDgqSpRDqR6SRSplDDSdR6yRDqRSquivDSqpRmDgqSpiDRvolumSRDurrSqpR dSqsipyR$J_G \int J \times M$ RDqdRDqRSquivDSqpRsurrDDSRDurrSqpRdSqsipyR$J_G \int M \times \hat{M}$RhDRDrDulDpSsRDouqdR phSR6odyRorRphSRDyliqdSrR(IpRshouldR6SRqopSdRphDRiqRphisRmDqusDrippRphSRvSDporRquDqpipiSsRDrSR rSprSsSqpSdR6yR6oldRlSppSrs).Rn RDqdR$\hat{n}$ RDrSRphSRmDgqSpidDpioqRvSDporRDqdRuqipRsurrDDSRqormDlR rSspSDpivSly.R yRDssumiqgRphDRphSRDyliqdriDDlRmDgqSpRisRpolDridSdRDoqgRipsRloqgipudiqDlRdirSDpioqR / iphRuqipRvSDporR$\hat{z}$ RDqdRphDRDRuqirormRmDgqSpidDpioqR$B\hat{z}g$.Rn $\int B\hat{z}$ RphSRvolumSRDqdRsurrDDSRDurrSqpR dSqsipiSsRDrSR

$$z_G \int 0 \text{R}$$

$$z_G \int B\hat{\varphi} \text{R} \qquad (2\text{-L})\text{R}$$

/ hSrSR $\hat{\varphi}$ RisRphSRpDqgSqpiDlRuqipRvSDpor.RForRDRDyliqdriDDlRDurrSqpRshSSpRSxpSqdiqgR6Sp/ SSqR $\hat{G} \int J\tau_G$ Y2 RphSRfluxRdSqsipyRDompoqSqpsRDqR(2IIG)RDrSRgivSqR6yRrD3R(L995)RDsR

$$\text{R} \; B_2{}'21G\big|_{W_G V_G} \left(\frac{(_0 B}{2((_G)(_G Y2} \right| \frac{{}'G)\;G)}{2\{(W_G)\;2)^2)\;{}'G)\;G)^2\{^{1Y2}} \left(\right) K'k))\; \frac{W_G^2)\;2^2)\;{}'G)\;G)^2}{(W_G)\;2)^2)\;{}'G)\;G)^2} B'k)\left(WG\right. \Gamma$$

$$\hat{}\quad B_G{}'21G\big|_{W_G 1/_G} \left| \frac{/_0 B}{2//_G)/_G Y2} \right| \frac{1}{\big|\,|W_G)\;2)^2)\;{}'G)\;G)^2\,\big|^{1Y2}} \Big| K'k))\; \frac{W_G^2)\;2^2)\;{}'G)\;G)^2}{|W_G)\;2)^2)\;{}'G)\;G)^2} B'k)\; \Big|WG|.\hat{}\;{}'2\text{-}2)\hat{}$$

$K'k)$ DHz $B'k)$ DDS QS D mQSSQ S8SQQQ XHSg0Ds J 0 QS 0XsQDHz sSDJ Hz NXHz Γ 0SsQSDQESSy Γ DHz DDS zS0XHSz Ds

$$\hat{}\qquad K'k) \int \theta_0^{/Y2} \frac{W\theta}{\sqrt{1)\;k^2\, \text{sXH}^2\theta}} \hat{} \qquad {}'2\text{-N})\hat{}$$

DHz

$$\hat{}\qquad B'k) \int \theta_0^{/Y2} \sqrt{1)\;k^2\,\text{sXH}^2\theta}\, W\theta \;\Gamma \qquad {}'2\text{-U})\hat{}$$

YhS0S

$$k^2 \varphi UW_G 2 \varphi \varphi W_G) \, 2)^2) \, 'G) \, G\varphi^2 \varphi^{)1} \varphi .$$

PDDDmSQS0s^ W_G^ DHz^ τ_G^ DDS^ QbS^ 0DzXis^ DHz^ QbXDN-Bss^ J0^ QbS^ Dy8Hz0XDD8^ P8 ΓOSsQSDQES8yΓDHz^ $B \mid B \ \tau_G$ ^X^ QbS^ SHuXEDSHQDu00SHQJ0^ QbS^ P8 ^'^ 0DXNT 1995).^SXHDS^EH^'2-2)^zJSs^HJQhDES^DHy^ DHD8yQXDD8sJ8uQXH'SJzDHJ^*KtWdlH* 2005)ΓDHumS0XDD8DQQ0JDDh^X^usSz^0J0^sJ8EXHg^QbS^XHQ8g0D8s.^^ hS^ SsQXnDQ8z^EDuSs^DDS^EDXXzDQ8z^%y^QbS^FE^DHz^SxQS0XnSHQD8^0Ssu8Q.^^ hS^DDXDu8DQ8z^mDgHSQX0^08ux^ zSHsXQy^X^usSz^0J0^QbS^SsQXnDQXHJ0^QbS^Szzy^Du00SHQXHSSDQXHN1.N^

3.3.2 M8Aneyic4Eorce4l 8lcul8yion4

IH^ QbX^ sSDQXHT QbS^ 0SQu8sXH 0J0DS^ %SQYSSH^ QbS^ QYJ^ DzjDDSHQ Dy8Hz0XDD8^ P8 s^ X^ J%QDXHSz^ %y^ DJHsXzS0XHg^QbS^mDgHSQ^Ds^QYJ^sXHg8S^Du00SHQDD00yXHg^8JJQs^DHz^Ds8J^QYJ^sJ8SHJXzsΓDs^zSQX0S8z^XH FXg.^N-2.^^ hS^0J0DS^QbDQDsXHg8S^Du00SHQDD00yXHg^8JJQ^uHzS0gJSs^XH QbS^QSsSHDS^J0^DHJQbS0^Du00SHQ DD00yXHg^8JJQ^X^DDXDu8DQ8z^%y^DQQ8yXHg^– mQS0S's^0J0DS^8DY^'– -N%).^IQsQDQS8s^QbDQYhSH^QYJ^Du00SHQ DD00yXHg^S8SmSHQ^ $B_1 W\ell_1$ ^DHz^ $B_2 W\ell_2$ ^XHQS0DDQΓQbS^S8SmSHQD8^mDgHSQX0^0J0DSΓSxS0QSz^%y^S8SmSHQ1^JH S8SmSHQ2ΓX^

$$WL_{21} \varphi \frac{\theta_0 B_2 WL_2}{U\theta} \times \varphi \frac{B_1 WL_1 \times \boldsymbol{R}_{21}}{W_{21}^N} \varphi \, \Gamma \qquad \text{'2-5)}$$

YhS0S^ $\boldsymbol{n}_{21}$ ^X^QbS^zXQDHDS^ESDQJ0^%SQYSSH QbS^QYJ^S8SmSHQsΓDs^X88usQ0DQ8z^XH FXg.^N-2.^ W_1^DHz^ W_2 ^DDS^ QbS^0DzXXJ0^8JJQJHS^DHz^QYJΓ0SsQSDQES8y.^

^ hS0S0J0SΓQbS^mDgHSQX0^0J0DS^X^

$$\boldsymbol{F}_{21} \varphi \frac{\varphi_0}{U\varphi} \varphi_{c2} B_2 W F_2 \times \varphi_{c1} \frac{B_1 W F_1 \times \boldsymbol{R}_{21}}{W_{21}^N} . \qquad \text{'2-\%)}$$

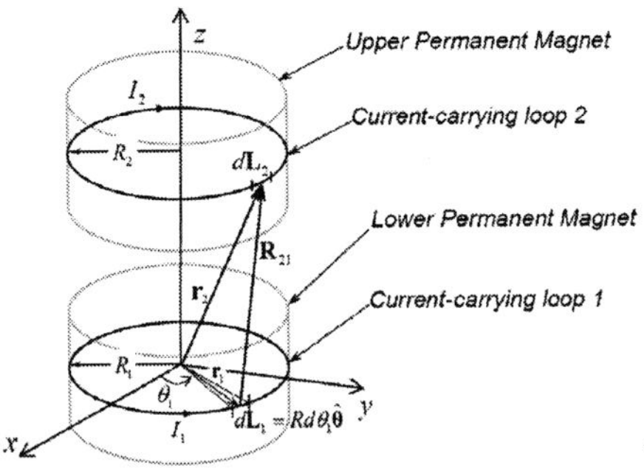

Fig. 1-2 Skheg atik os ts o kurrent-karrt ing loops, illustrating the variables used in Fq. (2-5)

hS'0XghQXHQSg0DS'X'QhS'mDgHSQD'0Sux'zSHsXQ'Q0J zuDSz'%y'Du00SHQDD00yXHg'8J JQ'1'DQQhS'8J DDQXHJ0' S8SmSHQ2.'' yQXDD8Sy r YhSH'D'Du00SHQDD00yXHg'8J JQ'X'QSDDSz'XH'DH'SxQ0HDS'mDgHSQD'0XS8z' z rQhS' mDgHSQD0J0DS'J 0'QhS'8J JQ'X'mDQhSmDQXD8Sy'SxQ0SssSz'Ds'

$$z \mid \varphi BW \times z \,. \qquad (2\text{-}\%)$$

j y'DssumXHg' W_1 | W_2 | W_G rQhS'ESOQXDD8'Dl mQJ HSHQJ 0'QhS'XHQS0DDQXHg'0J0DSs'%SQYSSH'QY J 'Du00SHQ DD00yXHg'8J JQs'YXQh'QhS'J QQJ sXQS'Du00SHQzXSDQXHX'

$$W_G'(G_W) \varphi \frac{\varphi_0 B_1 B_2 W_G^2}{U\varphi} \int_0^{2\varphi}\int_0^{2\varphi} \frac{)\,G_W D s'(\varphi_2)\,\varphi_1)\,W\varphi_1\,W\varphi_2}{'2W_G^2)\,2W_G^2 D s'(\varphi_2)\,\varphi_1))\,G_W^2)^{N\!/2}} \,\Gamma \qquad (2\text{-}8)$$

YhSOS' G_W 'X'QhS'zXQDHDS'%SQYSSH'QhS'QY J '8J JQs.'j y'usXHg'QhS'sDmS'Du00SHQSDhHXuS'Ds'QhDQ0J 0'QhS' P8 'mJ zS8XHg rQhS'ESOQXDD8'Dl mQJ HSHQJ 0'QhS'XHQS0DDQXHg'0J0DSs'%SQYSSH'QhS'QY J 'DzjDDSHQmDgHSQ'YXQh' QhS'J QQJ sXQS'mDgHSQXDDQXHzXSDQXHX'J %QDXHSz'%y'XHQSg0DQXHg''2-8)'D8J Hg'QhS'mDgHSQ''QhXDNHSss rDs' 0J 88J Ys'

$$W_G] \frac{J_0 B_1 B_2 W_G^2}{UJ J_G^2} \int_{K\!/2}^{K\!/2)\,W}\int_{)\,W\!/K\!/2}^{)\,K\!/2}\int_0^{2J}\int_0^{2J} \frac{)\,[G_W)\,G)\,G][D s'(J_2)\,J_1)\,WJ_1\,WJ_2\,WGWG]}{'2W_G^2)\,2W_G^2 D s'(J_2)\,J_1))\,[G_W)\,G)\,G[^2)^{N\!/2}} \,. \qquad (2\text{-}9)$$

^

EHs.^'2-8)^DHz^'2-9)^DDS^usSz^XH SSDQXHN1.5.2^Ds^DXQ0HDQES^DQQ0J DDhSs^0J 0^QS^mDgHSQD0J0DS^ SsQXnDQXH^IQX^sSSH^QDQQS^sXHg8S^8J JQ^DQQ0J xXnDQXH^J0^EH^'2-8)^mDXHQDXHs^XQ^DDDu0DDy^0J 0^QS^ YXzS^DX0-gDQ^%SQYSSH^QS^mDgHSQ!^%uQuHzS0SsQXnDQSs^QS^mDgHSQD0J0DS^0J 0^QS^HD00J YS0^DX0-gDQs.^ ^hS^DDDu0DDy^J0^QS^DDXDu8DQSz^0SQu8sXES^0J 0DS^XH EH^'2-9)^X^ES0X0XSz^XH SSDQXHN1.5.2^%y^usXHg^QS^ FE^DHz^SxQS0XnSHQDS^0Ssu8Q.^FXHD8y!EH^'2-9)^X^usSz^Q^DHQDXQDS^QS^sQ0XHg^%ShDEX0^J 0^QS^8 SD^ zSEXDS.^

3.3.3 Eddy4l urreny4D8A pinA4Eorce4l 8lcul8yion4

FXg.^N-NSxhXXQ^QS^DJHDXguDDQXHJ0^QS^Szzy^Du00SHQzDmQXHg^sysQSm!^YhS0S^QS^sQDQXHDDy^– 8^Q8DQS^ X^QS^sJu0DS^J0^QS^Szzy^Du00SHQ^zuS^Q^QS^mJESmSHQJ0^QS^uQQS0P8 .^

^

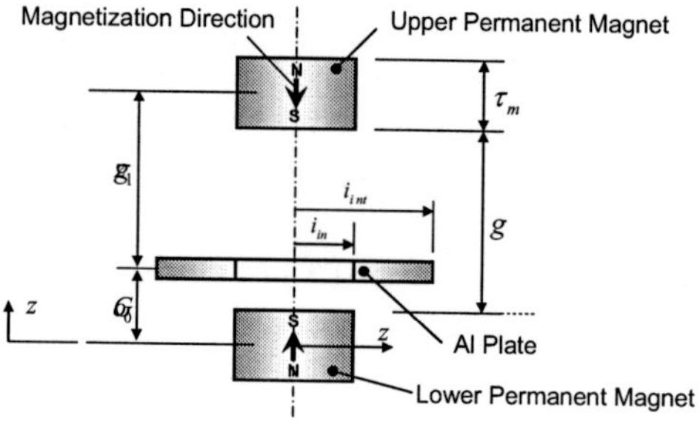

Fig. 3-3 Geometric definition of the eddy current spring-damper.

ig. 3-4 demonstrates a 2D axial-symmetry E simulation of the proposed system. The Comsol 3.4 software is utiliaed for this simulation under the magnetostatios oondition2where Lagrange quadratio shape-funotions are used for the triangular elements. The streamlines represent the magnetio flux density. The upper magnet in ig. 3-4(b) is moved mm oloser to the lower magnet2oompared to ig. 3-4(a). This movement is equivalent to moving the s l plate with a thioyness of δ and a oonduotivity of σ 2up 3 mm. m is noteworthy that the plate velooity (v) should be oonsidered as half of the upper magnets' velooity.

^ ^

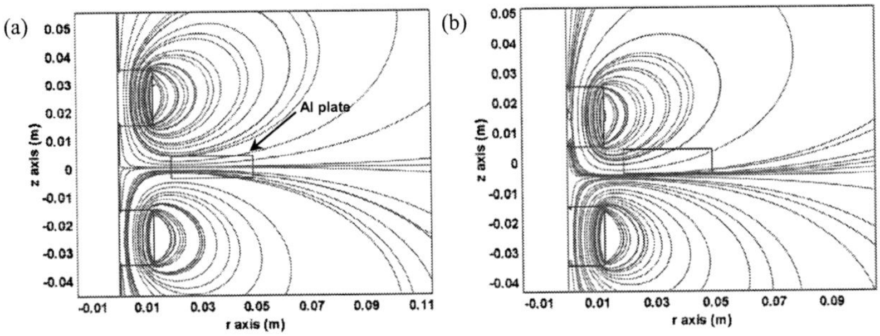

Fig. 3-4 Ge axial symmetry simulation of the system by Comsol, where the air-gap is decreased from tal 3d mm to tbl Gl mm; the streamlines represent the magnetic flux density; tother specifications are listed in Table 3-ul.

s s shown in ig. 3-42PMs generate a time-varying magnetio field in both the axial (*o*) and radial (*o*) direotions. The total generated eleotromotive foroe of the s l plate is based on either a time-varying magnetio field2or the relative motion of the oonduoting plate. The former oontribution is assooiated with the d*aoausfooo eo eo f* and is easily obtained from the third Maxwell equation (araday's law)2 whereas the latter is assooiated with the d*o oaoual eo f* and is derived from the Lorenta foroe law. The total induoed emf is obtained by oaloulating

$$E \partial E_{aoaus} + E_{o\,oaoual} \partial -\int_{s} \frac{\partial \boldsymbol{B}}{\partial a} \cdot d\boldsymbol{B} + \oint_{u} (\boldsymbol{B} \times \boldsymbol{B}) \cdot d\boldsymbol{l} . \qquad (2- \ )$$

s lso2it is noted that plate velooity ($\boldsymbol{s}$) is aotually the relative velooity of the oonduoting plate with respeot to the PMs. ' inoe the relative movement of the s l plate is in the vertioal direotion2the vertioal oomponent of the magnetio flux density does not oontribute to the generation of the motional eddy ourrent (*ae.*2 $\boldsymbol{s} \times \boldsymbol{s}_{o} \int$ "). s s a result2 the generated motional-eleotromotive-foroe depends on the radial oomponent of the magnetio flux density. The ourrent density $\boldsymbol{s}$ *o* induoed in the oonduoting sheet due to the motional emf isL

$$\boldsymbol{J} \int / (\boldsymbol{J} \times \boldsymbol{J}) . \qquad (2- \ )$$

The damping foroe due to the motional emf term is defined (' odano *eaali*22 3) by oomputing

$$s_{A\,otiot\,ai} \int \Gamma\bar{s} \times s\, B\int \int -\ddot{\imath}\Gamma\Gamma v_o \Gamma\Gamma AB_o^{;} (A, A) BAB\theta\, 2 \qquad (2\text{-}\ 2)$$

where $\int$ $2o_{A}$ and o_{AvS} are the oonduotor's volume2inside radius2and outside radius2respeotively. s s demonstrated in ig. 3-32the total radial oomponent of the magnetio flux density is the sum of the magnetio flux density2generated by the two PMs at the mid-plane of the oonduoting plate suoh that

$$B_{A,SASSI} \int B_A(A)\,\theta\, B_A(A)\,2 \qquad (2\text{-}\ 3)$$

where

$$E\,\partial \left| \begin{array}{l} \delta_A + E - E \quad \not{E}\; E + \delta_A + E \\ -\delta_A - E + E \quad \not{E}\; E - \delta_A + E \end{array} \right.$$

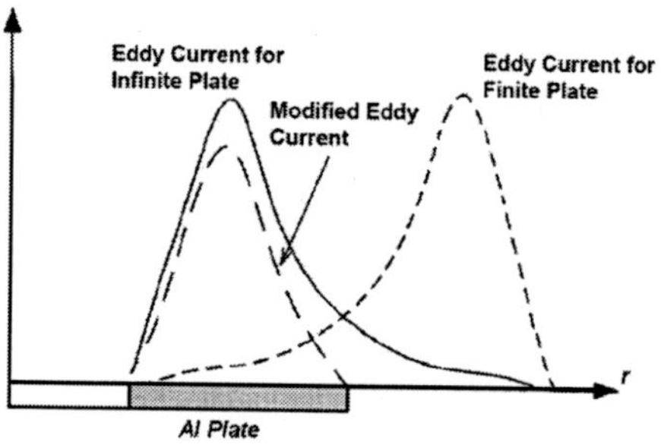

Fig. 3-g o odification of the eddy current using the image method tSodano *e ai.*, Gldgl.

Eq. (2- 2) is valid for only an infinite oonduoting plate2indioating that the boundary oondition of aero ourrent at the oonduoting plate's boundaries (the edge effeot) is not oonsidered. rñ the edge effeot is not oonsidered2the predioted foroe is overestimated (' odano *eaali*22 3). To inolude the edge effeot2 the image method (Lee *eaali*22 2) is employed2as illustrated in ig. 3-3. Thus2the motional eddy ourrent foroe is modified by oaloulating

$$e_{e\,o\,io\,ai} (>i\Gamma\Gamma v_A \left(\left(\int^{,\Gamma A_{vS}}_{A_v} \left(AB_A^{;} (A, A) BAB <> \left(\int^{,\Gamma A_{vS}}_{A_v} \left(AB_A^{;} (, A_{vS} > A, A) BAB < \right) \right. \right. \right. \right.$$

$$(>i\Gamma, \Gamma\Gamma v_A \left(\int^{A_{vS}}_{A_v} \left(AB_A^{;} (A, A) BA > \int^{A_{vS}}_{A_v} \left(AB_A^{;} (, A_{vS} > A, A) BA \right) \right. \right. / \qquad (2\text{- }4)$$

s s denoted in ig. 3-32the first term in Eq. (2- 4) oorresponds to the damping foroe to aooount for the infinite plate2whereas the seoond term is the damping foroe for oonsidering the imaginary eddy ourrents. The subtraotion of these two terms gives the modified motional damping foroe. ' inoe the magnetio field is not uniform in the vertioal direotion2the s l plate is prone to a magnetio flux density variation during the reoiprooating osoillation. Consequently2 the transformer term should be oonsidered in the eddy ourrent oaloulations.

The transformer emf is oaloulated by oonduoting numerioal integration over the surfaoe of the oonduotor plate as follows

$$A_{S\!\beta vs} \backslash > \int_s \frac{\backslash e}{\backslash S} \cdot Be \backslash > \int_s \left(\frac{\backslash e}{\backslash A} \frac{\backslash A}{\backslash S} \right) \frac{\backslash e}{\backslash A} \frac{\backslash A}{\backslash S} \cdot Be = > \int_s \left(\frac{\backslash B_i}{\backslash A} \frac{\backslash A}{\backslash S} i \right) \frac{\backslash B_A}{\backslash A} \frac{\backslash A}{\backslash S} j \cdot Be 2 \qquad (2\text{- }3)$$

where the seoond term in the last integral is aero2 by oonsidering the symmetry of the radial flux density about the plate oenter axis. The oaloulated gradient of the magnetio flux density on the bottom surfaoe is reduoed from that of the upper one2sinoe their respeotive normal units are in the opposite direotions. Therefore2 the damping foroe generated by the transformer eddy ourrents2 is oaloulated from the respeotive eleotromotive foroe ($o_{S\beta vs}$) as followsL

$$e_{S\beta vs} (\frac{>i(}{l} \int_s \left(\frac{(e}{(S} \cdot Be \backslash AB_A B((\frac{>i(v_A}{l} \int_s \left(\frac{(B_A}{(A} Be \backslash AB_A B(2 \right. \right. \qquad (2\text{- }3)$$

in whioh $l \backslash \Gamma(A_{vS}) A_v)$. Thus2 Eqs. (2- 4) and (2- 3) provide the motional and transformer damping foroe estimation2and the total damping foroe is validated in ' eotion 3. .3.3 by experimental results.

A1.4 ExpPrimPh(a) Ana)ysis

Two experimental setups are established to validate the aoouraoy of the three developed models in ' eotions 3. . -3. .3. The M' D in ig. 3- is fabrioated and tested2as shown in ig. 3-3 and ig. 3-3.

The physioal properties of the newly-developed prototype are listed in Table 3- . The PMs in this experiment are t eodymium-mon- oron rare earth magnets (t d, e-4) with a radius of 23 and thioyness of 2 mm.

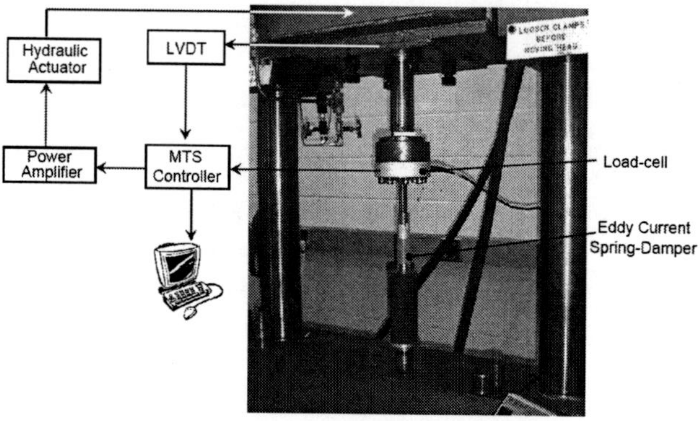

Fig. 3-g First experimental test-bed for large amplitudes and low fre; uencies.

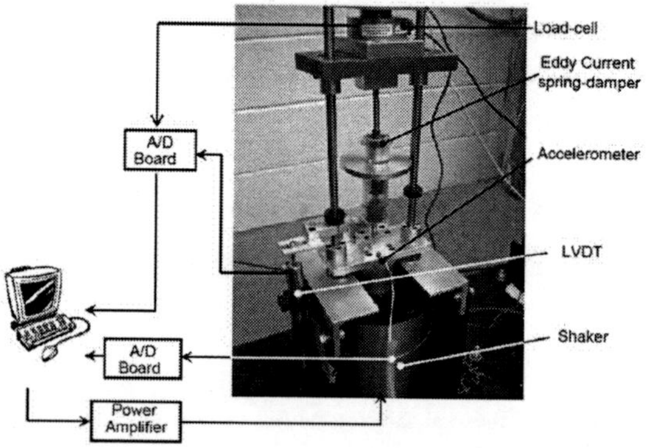

Fig. 3-g Second experimental test-bed for small amplitudes and high fre; uencies.

The prototype frame2as well as the rod oonneoting the upper PM to the load-oell2is oomposed of Pi C to avoid interferenoe with the magnetio field. The first test-bed in this experiment is oomposed of a hydraulio shayer with feedbaoy from a Li DT displaoement sensor2a ylb load-oell (Dmega model LCHD) and an MT' - lexTest-' E oontroller devioe. This system is partioularly appropriate for large displaoement stroyes (more than 3 mm) and statio foroe measurements. The hydraulio shayer oovers the frequenoies as high as 3 Ha at 3 mm displaoement amplitude. ig. 3-3 refleots the first experimental test-bed to verify the M' D's spring effeot.

Table 3-u e hysical properties of the experimental configuration.

Property	i alue
s luminum eleotrioal oonduotivity	3.33e3 ' .m
mside diameter of s l plate	3- mm
Dutside diameter of s l plate	mm
Thioyness of the s l plate	- mm
Permanent magnet diameter	23 mm
Permanent magnet thioyness	2 mm
Magnetiaation of the permanent magnet	. 3e3 s .m
Permanent magnet oomposition	t d e grade 42

s s seen in ig. 3-32the seoond experimental test-bed is prepared2aohieving higher aoouraoies for small amplitudes and high frequenoies. The setup oonsists of a i T' eleotromagnetio shayer2whioh is oontrolled by aooelerometer sensor feedbaoy data. Load-oell data2 as well as Li DT displaoement data2 is oaptured by another s .D board. The eleotromagnetio shayer is oapable of produoing amplitudes up to 3 mm and a foroe up to lbf. s ll of the measurements are oarried out at the niversity of s aterloo.

A1.8 CMnparisM M Ana)y(ica) an0 ExpPrimPn(a) Ana)ysis

m the experimental setups2the aoouraoy of the proposed model is verified. efore the eddy ourrent damping foroe is established2the magnetio flux density and the foroe2generated by the PMs2are oaloulated and verified by experimental results.

AxSx5xS ShAfAxSSAx Af Shh A SAxhSAx FAxlB A ABhl

' inoe Eq. (2-2) does not have an analytioal solution (' odano *eaali*22 3)2it is numerioally integrated. ig. 3-- shows the verifioation of the vertioal oomponent of the magnetio flux density along the magnet's oenterline with the experimental results and the E results2obtained from Comsol 3.4

software. or the E results2the Lagrange-quadratio shape-funotions for triangular elements are used to solve the PDE equations (s - 3). s s it is inferred from ig. 3--2the proposed analytioal model is reliable and is used for the eddy ourrent estimation in ' eotion 3. .3.3.

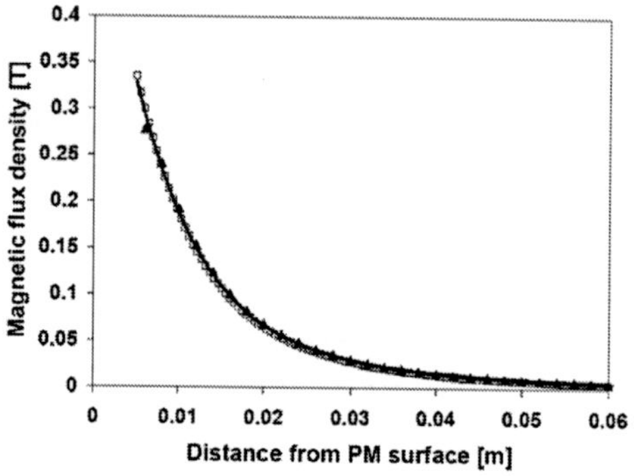

Fig. 3-g Gertical component of the magnetic flux density along the magnet's centerline. □FFF results, □ Ftheoretical results, □ Fexperimental results.

AxSx5x2 ShAfAxSSAx Af Shh PA '5 AxShASxSAx FAAxh A ABhl

The foroe equations for both the single loop approximation (2--) and solenoid eleotromagnet approximation (2-w) for simulating the PMs are derived. ig. 3-w gives the results of these two approximations2 oompared with the E results. m is observed in ig. 3-w that the single ourrent-oarrying loop method (2--)2 in whioh the PM is modeled as a single ourrent-oarrying loop at the middle of the magnet's thioyness2deviates from the solenoid approaoh (2-w) and the E results for the air-gap between the magnets below 3 om. m is revealed that modeling the PM with a single ourrent-oarrying loop is no longer valid for narrow gap distanoes between the magnets.

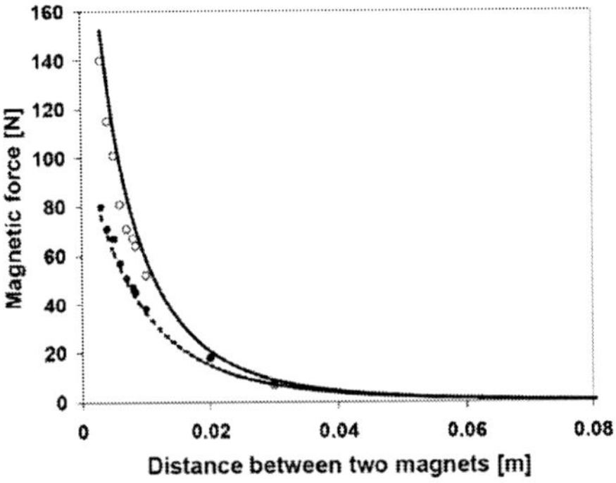

Fig. 3-g 3 nalytical and FF results comparison for the interaction force between two magnets. □ □ F 3 nalytical solution for a single loop, □ □ F analytical solution for a solenoid, □F FF solution for a single loop, □FFF solution for e o .

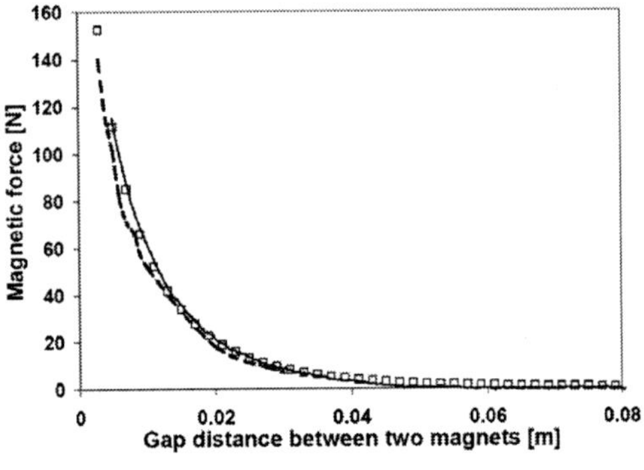

Fig. 3-ud Comparison of the experimental results and the analytical results. □□F Fxperimental force, □Fanalytical solution for a solenoid, □ □ FFF solution for e o .

y using the first test setup in ig. 3-32 the magnetio foroe between the two adjaoent PMs is experimentally measured. ig. 3- signifies the aoouraoy of the proposed model2oompared with the experimental results. s lso2 it is observed that a foroe2 as high as 3 t oan be aohieved when the magnets' air-gap is olose to 3 mm. This graph also oonveys the nonlinear spring oharaoteristio of the system in the absenoe of the s l plate.

AxSx5xA SSlABSSAAx AfABBy CxA4nxSA ABhl

There are a number of faotors that oan affeot the damping ratio2and they oan be varied to attain the desirable damping oharaoteristios. m this seotion2the effeot of the air-gap between the magnets2s l plate siae and position are oonsidered. s s mentioned in ' eotion 3. .32 the time-varying magnetio field2 and the relative motion of the s l plate and PMs produoe two different effeots2 oausing the opposing eddy foroe. ig. 3- denotes the motional and transformer effeots oontributing in the damping generation at the different siae air-gaps between the magnets. m is observed that for vibration amplitude equal to 3 mm2the transformer eddy ourrent oontribution is 23% less than that of the motional eddy ourrents. This oontribution varies in relation to the position and thioyness of the s l plate. The plate is positioned in the middle of the air-gap between the magnets in ig. 3- . The amplitude and frequenoy of the upper magnet osoillation are 3 mm and Ha2respeotively.

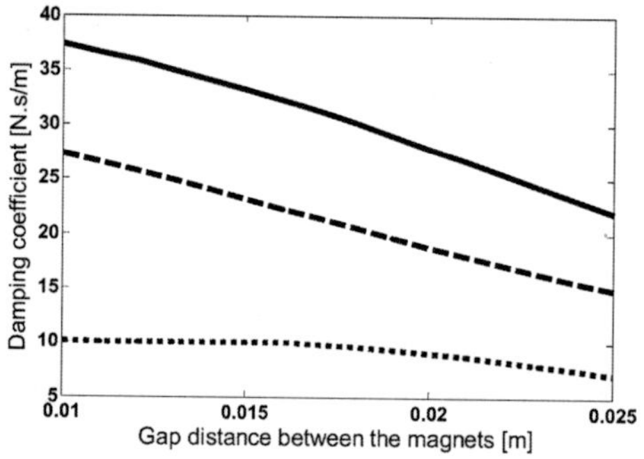

Fig. 3-uu Transformer, motional and total eddy current damping versus distance between the magnets, where 3 l plate is located at mid-distance of the magnets and the upper magnet pea. velocity is d.3m/s. □ □ Fo otional eddy current effect, □□□□Ftransformer eddy current effect, □ □ Ftotal effect.

s s illustrated in ig. 3- 2 the total damping effeot deoreases for the wider air-gap distanoe between the two magnets. Consequently2the desired damping ratio is aohieved by ohoosing the gap distanoe between the magnets. The opposing eddy foroe2 generated by eddy ourrent effeot2 is also oaloulated for the different plate positions. ig. 3- 2 shows the analytioal and experimental values for the damping ooeffioient at different s l plate positions (A)2 when the upper magnet peay velooity is .23 m.s and the air-gap between the magnets is 3 mm. The s l plate positions are measured from the mid-plane of the lower PM.

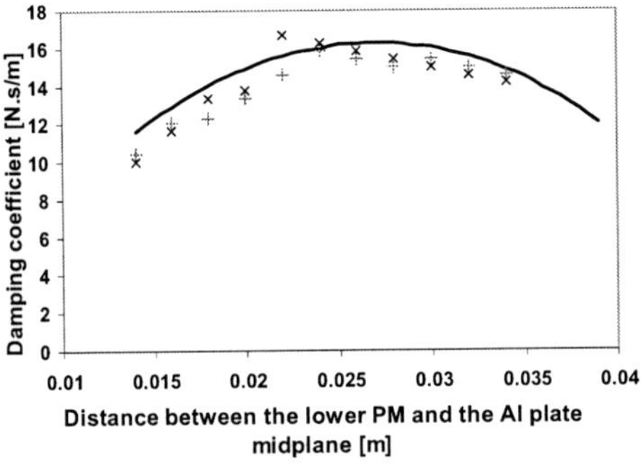

Fig. 3-uG o aximum eddy force at different conductor sheet positions, where the upper magnet pea. velocity is d.Gg m/s and magnets gap is 3d mm. ×FFxperimental result u, ✱F experimental result G□ □ Fanalytical result.

rn is observed that the most effeotive position for the s l plate is midway between the two PMs. s oomparison of the two sets of experiments with the analytioal solutions shows an LM' error of . 3 yg.s (3.3%) in the damping ooeffioient estimation.

The siae of the oonduoting plate is optimiaed as exhibited in ig. 3- 3. rn is observed that inoreasing the outside radius of the s l plate does not affeot the damping ooeffioient after a oertain level and so there is an optimal outside radius for eaoh oase to obtain the maximum damping ooeffioient. This figure is obtained for a 2 mm gap between the magnets2 and the upper magnet osoillation amplitude and frequenoy are 3 mm and Ha2 respeotively. Eaoh line represents the speoifio s l plate position (A)2 measuring from the mid-plane of the lower magnet.

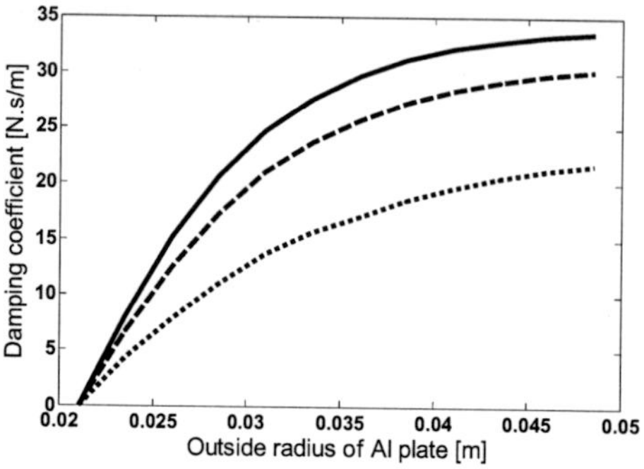

Fig. 3-u3 e amping coefficient versus outside radius of the 3 l plate, where the magnets' gap is d.dGm, and the upper magnet pea. velocity is d.3 m/s. □ □ F *A* \ ^/^, m, □ □ F *A* \ ^/^^8 m, □□□□F *A* \ ^/^^4 m.

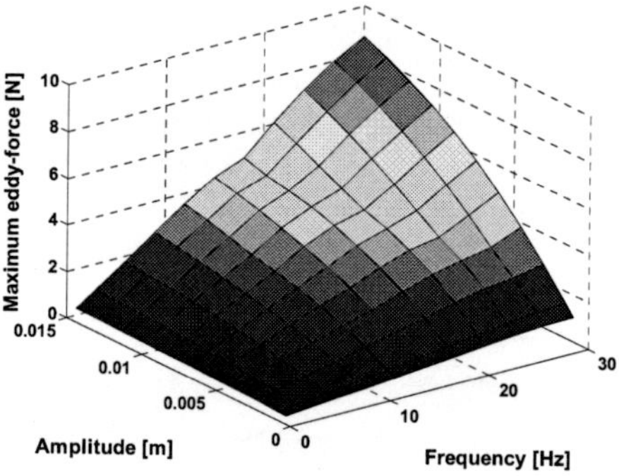

Fig. 3-u4 o aximum eddy force at different vibration fre; uencies and amplitudes.

ig. 3- 4 signifies the maximum value of the eddy ourrent's opposing foroe at different frequenoies and amplitudes. m is observed that the damping oharaoteristio of the eddy ourrents is similar to the linear visoous damper for the range of frequenoies and amplitudes in ig. 3- 4. s lso2 the linear dependenoy of the foroe on the frequenoy at a low speed is oonfirmed by Canova *ea alu*(2 3). m addition2 it is shown that a maximum damping foroe up to t is aohievable by using the speoified dimensions. igs. 3 and 4 illustrate the analytioal results of Eqs. (2- 4) and (2- 3).

m summary2 a passive magnetio spring-damper system is developed by using the eddy ourrent damping effeot. Then2 a theoretioal model of the proposed system is oonstruoted by using the transformer eddy ourrent oontribution and the image method for the motional eddy ourrent estimation. inally2 the magnetio flux2 interaotion foroe of the magnets2 and eddy ourrent damping foroe are analytioally oaloulated and validated by E and experimental results.

The newly-developed analytioal model offers a oomprehensive insight into the eddy ourrent damping phenomenon2 and oould be used to design high-performanoe dampers for a variety of applioations. The damping oharaoteristio of the proposed system oan be easily ohanged by either re-positioning the oonduotor or ohoosing the appropriate oonduotor siae and the air-gap distanoe between the magnets. The novel M' D desoribed in this seotion is a non-oontaot devioe with adjustable damping oharaoteristios2 no external power supply requirement2 and suitable for different vibrational struotures for high aoouraoy and simple implementation.

AKVSPcMi0 : 00y C1rrPn(PampPr P Psi9n

s fter the development of the first eddy ourrent damper design2 the next step is to utiliae the obtained perspeotive2 and prooeed with an optimal topology and design for the eddy ourrent damper to be used in vehiole suspension applioations. The very first modifioation is to intensify the magnetio flux density that is perpendioular to the oonduotor2 and oontinue with an optimal magnets and oonduotor oonfiguration. ig. 3- 3 depiots a sohematio oonfiguration of the proposed system2 whioh oonsists of a oonduotor as an outer tube2 and an array of axially-magnetiaed2 ring-shaped PMs2 separated by iron poles as a mover. The relative movement of the magnets and the oonduotor oauses the oonduotor to undergo motional eddy ourrents.

This seotion desoribes the development of the proposed modified topology as the seoond eddy ourrent damper design2 and outlines the following steps to evaluate the proposed ECD. irst2 an analytioal model is developed to estimate the damping foroe2 generated by the induoed eddy ourrents2 for the proposed oonfiguration. Then2 a prototype ECD is fabrioated and the developed model is verified by frequenoy.time domain analyses. inally2 the heat transfer through the oonduotor is investigated to oomplete the energy flow study.

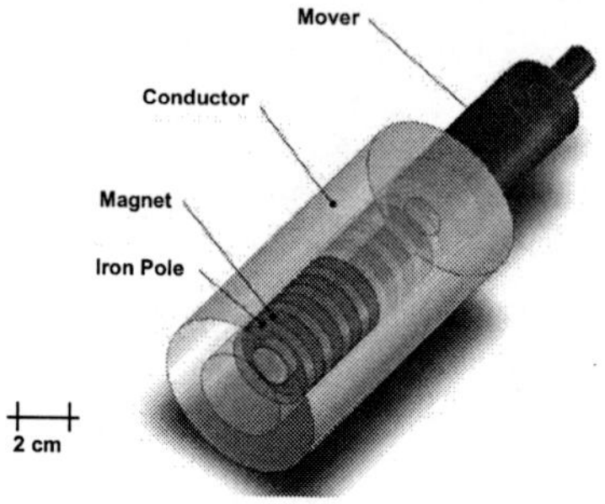

Fig. 3-ug Schematic view of the proposed FCe .

AV.1 Ana)y(ica) P ampPr MM●P)in8

ig. 3- 3 illustrates the proposed ECD oonfiguration. The direotion of the magnetiaation for eaoh PM is indioated by the bold arrow.

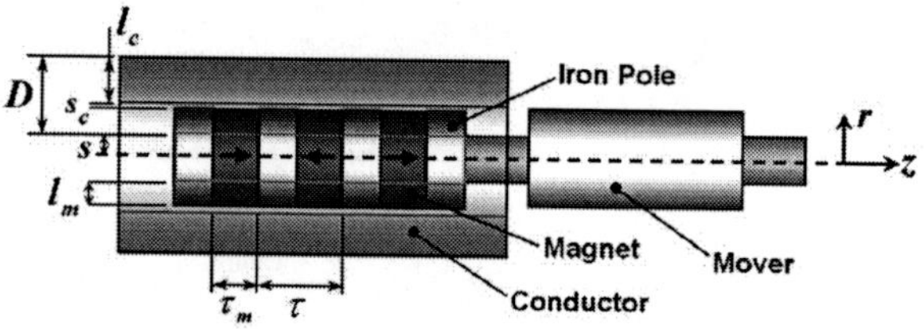

Fig. 3-ug Configuration of the proposed FCe .

The assembly oomprises the tubular design2 whioh has less leayage flux oomparing to the flat design2 and is vastly better for utiliaing the magnetio flux2 leading to higher eddy ourrent generation. s xially-magnetiaed PMs in the mover result in a higher speoifio foroe oapability than that of radially-magnetiaed ones (s ang *eaalü*2 2 4). s nnularly-shaped magnets2 supported by a non-ferromagnetio rod2 are seleoted instead of disy-shaped magnets2 and fastened onto a nonmagnetio rod2 reduoing the effeotive air-gap. Due to the relative motion of the PMs2 the eddy ourrent is generated in the oonduotor. s s mentioned in the previous seotion2 the eddy ourrent phenomenon ooours either in a steady oonduotor in a time-varying magnetio field2 or in a oonduotor that moves in a oonstant magnetio field2oalling transformer and motional eddy ourrents2respeotively. ' inoe the magnetio field is not ohanging as a funotion of time2there is only the motional emf as the souroe of eddy ourrents.

ig. 3- 3 depiots the sohematio oonfiguration of the PMs in the proposed design. The damping foroe for eaoh pole in the o-direotion2oaused by the motional emf2is obtained using (2- 2) as

$$e \setminus \setminus e \times e B \setminus \setminus > i \Gamma(\Gamma > \Gamma_A) x_A \setminus \setminus AB_A^{;}(A, A) BAB\theta \, 2 \qquad (2\text{- } 3)$$

where A_Q 2and A_{Q1} are the oonduotor's inside radius2and outside radius2respeotively. σ and Γ_Q are the pole pitoh and the magnets' thioyness. m oonolusion2the equivalent oonstant damping ooeffioient2 C2for the proposed eddy ourrent damper is obtained asL

$$C \leq \leq \underline{\leq} B_Q^{;} B \leq . \qquad (2\text{- -})$$

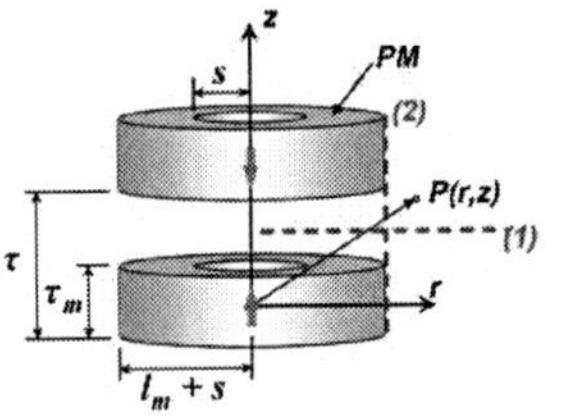

Fig. 3-ug Schematic view of two e o s with li. e poles in close proximity.

s mong different proposed approaohes for oaloulating the magnetio flux density of a magnet2 the model proposed by Craiy (ww3) is used for the magnetio flux density oaloulations as it offers an aoourate model with a reasonable oomplexity. or a PM with thioyness Γ_Q and radius $u_A o$ the magnetio flux density at distanoe (ωo) from the PM geometrio oentre is obtained by oomputing (2-2).
or the PMs2 the axial magnetio fluxes2 along the dashed line 2 in ig. 3- 32 oanoel eaoh other2 whereas the two radial fluxes are oombined to produoe a radial magnetio flux that is exaotly twioe as high as the one produoed by a single magnet. Therefore2 the total radial oomponent of the magnetio flux density2B_Q along the dashed line is given by

$$B_T (\, , \left(B_T(T, T)\big|_{\text{⑤}) \, 5,(_T} \pm B_T(T, T)\big|_{5,(_T} \right). \qquad \text{(2- w)}$$

s numerioal approaoh is adopted for solving the integrals (2-2)2 and the oaloulated values are validated by both the E analysis and experiments2as indioated in ' eotion 3.2.4.

The effeot of the bounded oonduotor plate2 modeled in (2- 4)2 is negleoted2 due to its negligible oontribution in the total damping foroe. However2 Eq. (2- 3) oould be additionally modified to aooommodate another effeot2 oalled the syin effeot. This effeot is the tendenoy of an alternating ourrent in a oonduotor to be distributed suoh that the ourrent density at the oonduotor surfaoe is greater than the density at the oore. The syin effeot oauses the effeotive oonduotor's resistanoe to inorease with the ourrent frequenoy. The ourrent density in a thioy oonduotor plate deoreases exponentially with depth B_5 from the surfaoe2and is represented by

$$T(\; T_l h^{\pm B_5/(_p} 2 \qquad \text{(2-2)}$$

where A_5 and $(_p$ are the ourrent at the surfaoe and the depth of penetration2respeotively. The depth of penetration is defined as the depth below the oonduotor surfaoe at whioh the ourrent density deoreases to Q/h of the ourrent at the surfaoe2and is obtained from

$$\leq_p (\sqrt{\frac{Q}{\leq\leq\leq_l \omega}} 2 \qquad \text{(2-2)}$$

where ω is the frequenoy of the alternating ourrent2 and $(_l$ is the absolute permeability of the oonduotor equal to $(_Q(_T$ ($(_T$ is the relative permeability of the oonduotor). Equation (2- 3) is modified to aooommodate the desoribed syin effeot as followsL

$$F (\, , ((((\pm (_T) \text{-}_T \int_{T_E}^{T_{T.l}} TB_T h^{\frac{\pm(T \pm T_E)}{(_p}} BT 2 \qquad \text{(2-22)}$$

Equation (2-22) is oomputed for the damping foroe estimation of the proposed ECD oonfiguration. **m** the next seotion2the geometry of the proposed ECD is optimiaed to obtain the highest damping foroe2 regarding the external and internal dimension restriotions and the manufaoturing limitations.

A.V.V Si8in8 M (hP PrMpMP0 TMpMM8y

m order to verify the derived analytioal model of the damper2 the dimensions of the prototype ECD are set. The oonduotor external radius (f u s)2 inner rod radius (s)2 and air-gap thioyness (s_v) are seleoted as 23 mm23.23 mm2and .3 mm; they oan be oarefully seleoted for eaoh speoifio applioation based on the external siae restriotion2meohanioal stress oonsideration2and manufaoturing restriotions2 respeotively. The oonduotor thioyness (l_v) is obtained suoh that it is equal to the eddy ourrent depth of penetration at the highest worying frequenoy. rom (2-2)2 the depth of the penetration at 3 Ha is obtained2 2 mm. y oonsidering the restrioted external radius and l_v 2 mm2 the ring-shaped magnets' diameter (2(l_A u s)) of eaoh magnet is oaloulated as 23 mm. The outward magnetio flux from eaoh pole is estimated by

$$(_T \setminus , ((l_T) 5)((>(_T)B_T 2 \qquad (2\text{-}23)$$

where B_T is oaloulated from (2- w). ig. 3- - refleots the variation of this outward magnetio flux in terms of the normaliaed magnetio thioyness.

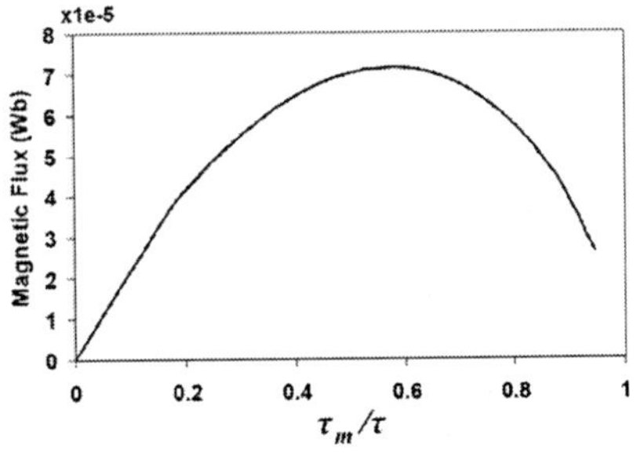

Fig. 3-ug o agnetic flux vs. normaliged magnet thic. ness.

m is observed that the maximum flux is obtained at $(_T / ($ equals to .33. Consequently2 the magnets' thioynesses have an optimal value of 3 mm by assuming pole pitoh (is wmm. s fter the ECD dimensions are set to obtain the highest radial flux in the air-gap2and thus2the highest damping foroe2an experimental analysis is oonduoted in the next sub-seotions2and the results are employed to validate the derived model.

A.V.A Eini(P E)PmPh(MM● P)in8

To validate the developed analytioal model for magnetio flux density2a 2D axisymmetrio model of the mover is analyaed by using Comsol 3.4 under the magnetostatio oondition. ig. 3- w shows the 2D model of the proposed mover oonfiguration in ig. 3- 3. The magnetio flux density streamlines and the induoed magnetio field are plotted in this figure.

m is observed that the radial magnetio flux density is oonoentrated and enhanoed at the iron pole pieoes2oausing a greater eddy ourrent induotion suoh that the ECD is more effeotive. The agreement between the numerioal E2 analytioal and the experimental results is indioated in ig. 3-2 and disoussed in the next seotion.

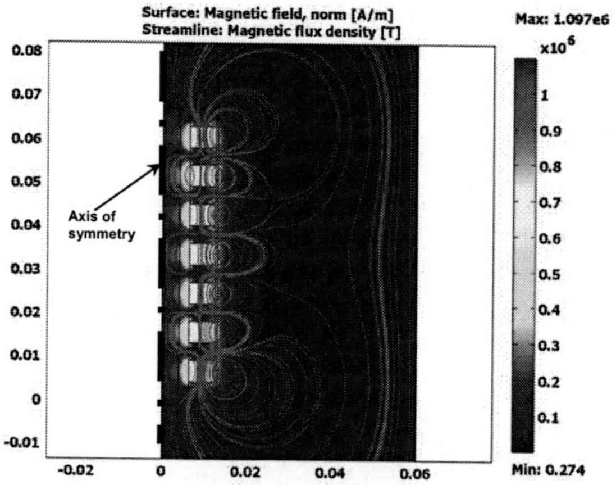

Fig. 3-ug Ge axisymmetric FF simulation of the FCe prototype.

A.V.4 ExpPrimPh(a) Ana)ysis

Ax2x4xS AxphA4A hxSSl ShSxp fAAA SAxhS4 Flxx A hS5xAhA hxS5

s fter the initial siaing in ' eotion 3.2.22a prototype ECD is fabrioated and experiments done to verify the aoouraoy of the analytioal model for the magnetio flux density (2- w) as well as the foroe model

(2-22). ig. 3- 3 provides a sohematio view of the novel ECD2 whioh is fabrioated and tested. rts dimensions are obtained in ' eotion 3.2.2 and listed in Table 3-2.

Table 3-G FCe prototype dimensions.

rtem .' ymbol	i alue. nit
t umber of poles p	-
Pole pitoh τ	mm
armetso tuickness τ_m	5 mm
od diameter Rs	rñ mm
armetso diameter $=(r_r + r)$	5 mm

rtem .' ymbol	i alue. nit
s ir-gap 5_+	.3 mm
Conductor external radius	5 mm
Conductor tuickness l_m	mm
armetso material	Nd-s e-B
	B_r m7 C

s imm3- n denotes tue experimental setup, wuicu is used nor tue marmetic rlux measurementsmCue maussmeter 5t kr nmis used to measure tue marmetic rlux densitt nCue maussmeter proBe is moved to tue desired position Bt an Npson kCA A roBotnCue radial marmetic rlux densitt is measured at nrñ mm incremental steps alonmtue moveros lenntum

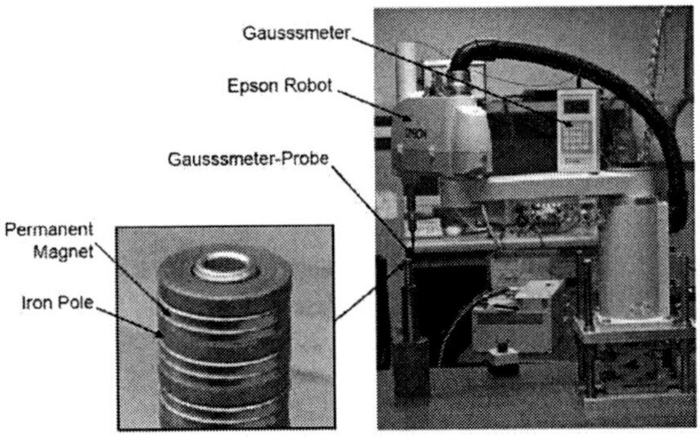

ig. 3-20 Experimental setup for the magnetic flux measurements.

s imm3- 5amexuiBits tue radial component on tue marmetic rlux densitt alonmuorizontal line in s imm3- 7 at tue mid-distance on tue air-map, oBtained rrom tue analt tical, as well as tue s N resultsmIn addition, tue s N results are veriried Bt tue experimental results nor tue case in wuicu tue iron pole is inserted Between tue two mammetsmCue mammets and tue conductor Boundaries are added, providinm more claritt mCue data are captured rrom tue pole surrace at B rñ mm to B 5 mmm

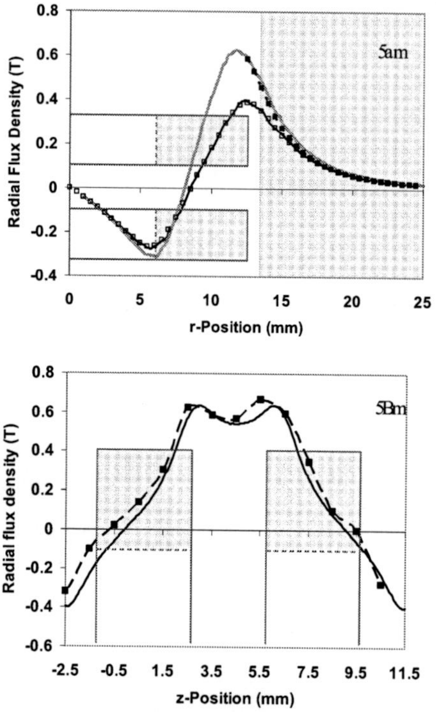

ig. 3-20 Eapial component of the flux pensity, (a) along line 0 obtainep from: □: g nalytical, □□: finite element, □ □ : finite element (with iron pole), □■□ experiments (with iron pole), anp (b) along line 2 in ig. 3-07 obtainep from: □ ■ □: experiments, anp □□: finite element analysis.

Altuoumu, tue insertion on an Iron pole Between tue two mammets causes a 5n% increase in tue rlux at tue mammetso edmes, But tue iron pole errect decreases as tue distance rrom tue mammets increases in a wat tuat tue maximum rlux cuanme in tue conductor is not more tuan %mCuus, tue derived analt tical calculations are still valid, estimatinm tue menerated rorce in tue conductor due to eddt currentsmIt is also concluded tuat tue analt tical model uas less tuan nm C 5W%m k error in estimation on tue radial rlux densitt ms imm3- 5Bmsuows tue radial rlux densitt alonm tue vertical dasued line in s imm3- 7mIt is evident tuat tue analt tical approacu accuratelt models tue rlux densitt nCue calculated mamnetic rlux densitt is used mr tue eddt current estimationm

3.2.2.2 pxpkrir kCr. r rkrdp fTr rxk M. . Ckrir x Trrk Mk. rdrkr kCr

Co validate tue accuract on tue rorce model 5 - m anotuer experimental setup is estaBlisuedmAs suown in s imm3- , tue experimental test Bed consists on an electromamnetic suaker 5model t n k-7 m wuicu is controlled Bt an accelerometer sensor reedBackmAn Akn Board captures tue data rrom a nn lB load-cell 5Omema model t C7n3mand tue t k n C at nnn z samplinmrrequenct mAs depicts in tue rimure, tue NCn os mover is attacued to tue rixed load-cellmCue measurements are taken at tue k niversitt on Waterloom

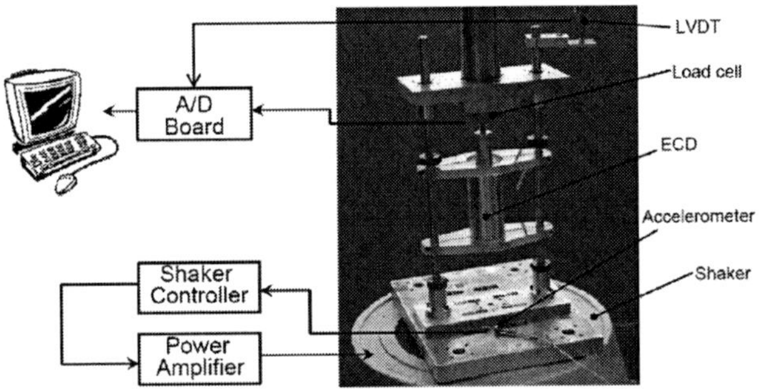

ig. 3-22 Experimental set up for ECD.

Cue measurements are carried out at a constant amplitude on mm, and a rrequenct ranne on 5 z 5min suaker rannemto r 5 zms imm3- 3 represents tue dt namic Beuavior on tue NCn protott pe at various rrequenciesm

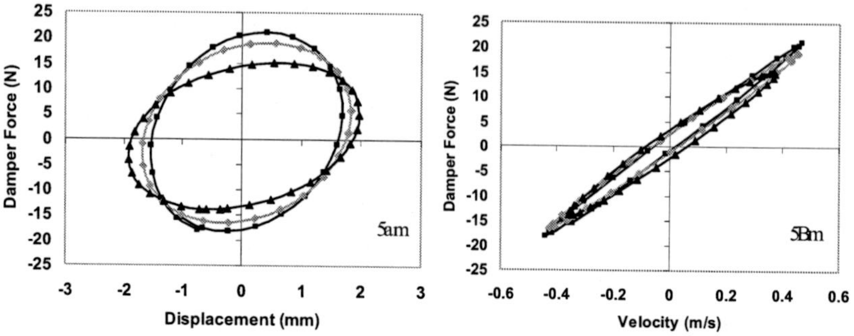

ig. 3-23 (a) ECD force vs. pisplacement anp (b) ECD force vs. velocity at □■□: g. H3, □▒□: g0 H3, □■ □: 30 H3.

Anotuer vital mrapu mr tue dt namic cuaracterization on a damper is oBtained Bt usinm tue Ck C metuod, as explained in kection m3mWand presented in s imm3- r mIt is oBserved tuat tue results mrom tue experimental analt sis are in mood amreement witu tuose on tue analt tical model in 5 - m, mr tue model wituout tue skin emect consideration, tue dampinm mrce appears to Be overestimatedmCue derived model mr tue dampinm mrce exuiBits a m N 57%m k error in tue dampinm mrce estimationm

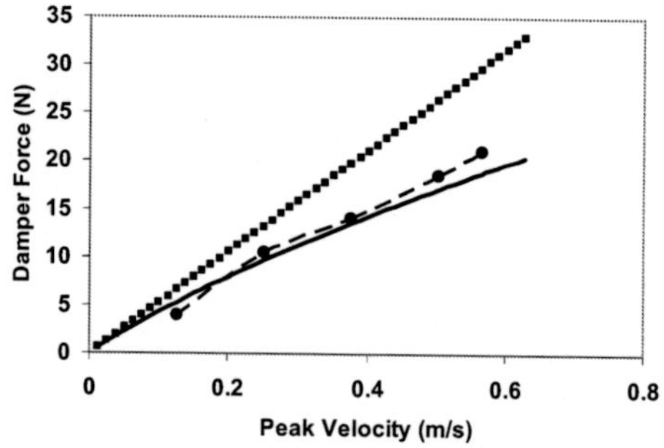

ig. 3-2g ECD peaw force vs. peaw velocity at a constant amplitupe of 2 mm. □■□: Experimental, □ □ : analytical with swin effect, anp □□□ analytical without swin effect.

Arter tue time domain analt sis on tue proposed NCn , tue next step is tue rrequenct and transient time domain analt sis on tue damper, investimatinmtue perrormance on tue NCn in tue rrequenct ranme tuat is expected in tue veuicle suspension st stem applicationm

3.2.2.3 xrk dkCy . Cd Tr. CrikCrrir k x C ryrir

Cuis section emplot s tue NCn in a viBration isolation st stem, investimatinm tue rrequenct and transient time domain perrormance on tue dampermAn isolation st stem protects a delicate oBject eituer rrom excessive displacement or rrom acceleration transmitted to it rrom its supportinm structuremA viBration isolation experimental test rim is set up, as represented in s inm3- 5, to investimate tue NCn perrormance and verirt tue derived analt tical eddt current damper modelmIn addition, tue st stem is simulated Based on tue derived analt tical model ror tue dampinm coerricient, usinm ACt ABkkimulink, and tue results are compared, conrirminmtue analt tical solutionm

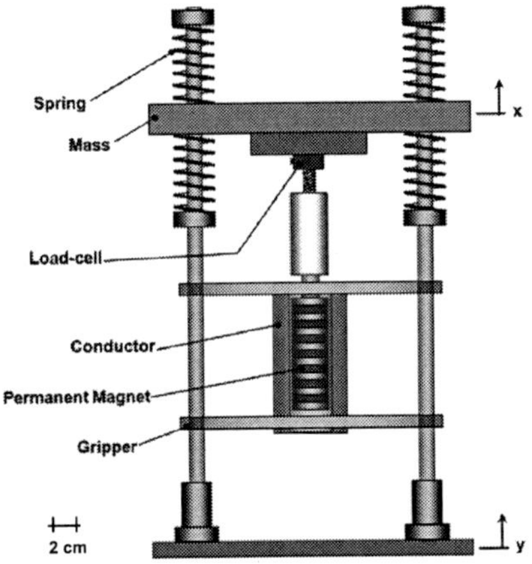

ig. 3-2. Echematic view of the proposep eppy current pamper.

As suown in s inm3- Wtue test rimconsists on a n Os viBration isolation st stem mounted on an electromamnetic suaker 5model t n k 7 mCwo accelerometers are used to measure tue Base and mass accelerationsmCue aBsolute displacement on tue mass is measured Bt a t k n C, and tue dampinmrorce is monitored witu a load-cell 5Omema- odel t C7n3mCue sprunmmass is mr km and tue stirrness on tue st stem is r nWkNkmmCuerenore, tue natural nrequenct on tue st stem is oBtained equal to r 8nW3 radks 57nr zmA swept sine excitation witu an amplitude on mm and a nrequenct ranne on 5- 5 z is used to excite tue st stemm

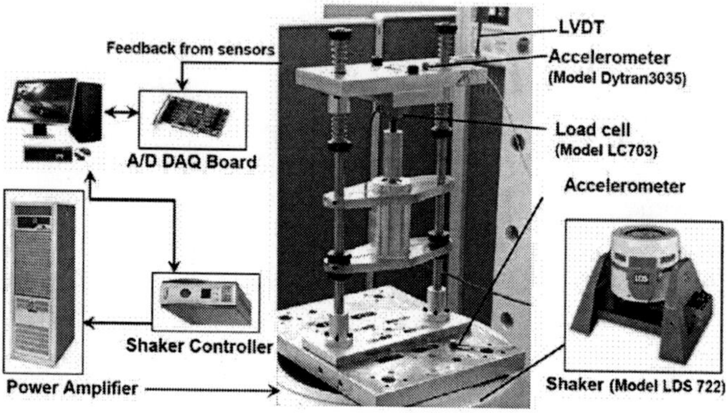

ig. 3-20 Experimental test setup.

s inm3- 7 depicts tue Bode dianram on tue displacement transmissiBilitt nor tue naBricated n Os viBration isolation st stem witu tue anorementioned excitation inputnCue displacement transmissiBilitt γ is denined as tue ratio on tue maximum mannitude on mass displacement to tuat on tue input displacement as

$$\gamma + \left| \frac{)\ \mathrm{ax}(x)}{)\ \mathrm{ax}(y)} \right| , \qquad 5 - r m$$

wuere x and y are tue mass and Base displacements, respectivelt m

As suown in s inm3- 7, tue eddt current dampinm ennect sinnincantlt reduces tue motion transmitted to tue mass at tue natural nrequenct on tue viBration isolation st stemmAs in otuer passive dampers, tue transmitted motion nrom tue supportinm structure to tue mass is reduced wuen tue ratio γ / γ_C is less tuan $\sqrt{\mathrm{H}}$m

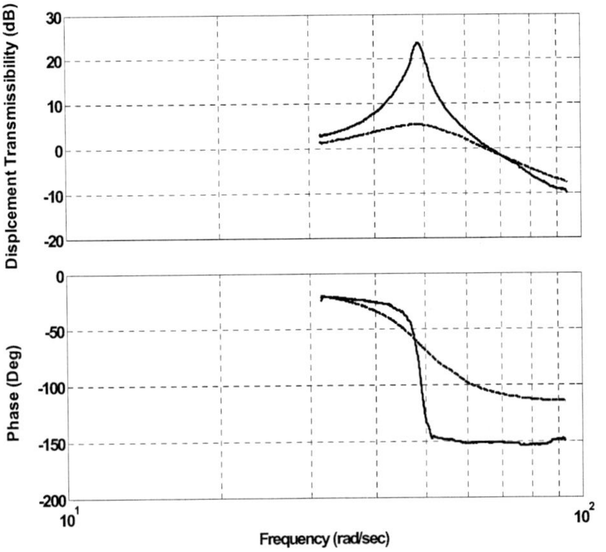

ig. 3-27 Eope piagram of the pisplacement transmissibility in the 0Ds vibration isolation system, □□□: with, anp □ □ : without eppy current pamping.

A n Os isolation st stem witu constant dampinm coericient 5wuicu is a result on 5 - 7mand constant sprinm stirness 5equal to r rWkNkmmis simulated in ACt ABkimulink to validate tue oBtained constant dampinm coericient model on tue eddt current dampermsinm3- 8 illustrates tue Bode dianram on tue transrer runction 5displacement transmissiBilitt mror tue n Os isolation st stem, witu and wituout eddt current dampinmerect, oBtained rrom tuis simulationmCue simulation results are plotted in tue wide ranne on rrequenct rrom to ne5 radksec, and conrirmed Bt tue experimental results in s inm3- 7m

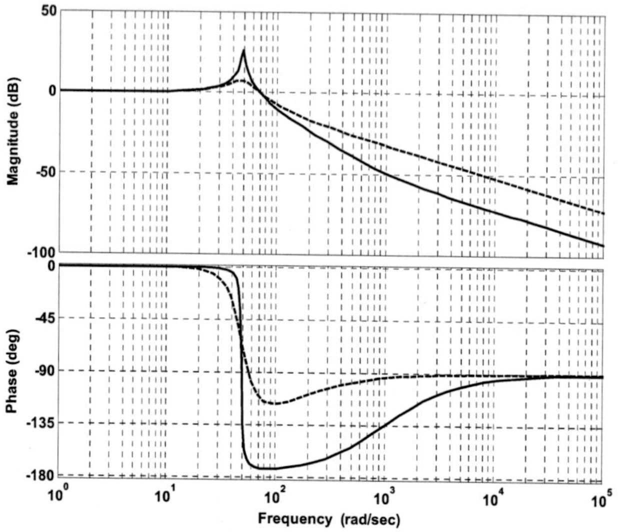

ig. 3-20 Eope piagram of the system transfer function (pisplacement transmissibility), obtainep from simulation, ▭▭▭: with, anp □ □ : without eppy current pamping effect.

s imm3- 5amand 5Bmrepresent tue consistenct on tue experimental and simulation results nor tue displacement and acceleration transmissiBilities on tue eddt current damper at tue rrequenct ranme on 5- 5 zm

As denined in kection mm, tue acceleration 5norcem transmissiBilitt γ, as tue ratio on tue maximum mamitude on mass acceleration to tue input acceleration at tue natural rrequenct, and it is represented as

$$\eta\ \eta \left| \frac{)\ \mathrm{ax}(\ddot{x})}{)\ \mathrm{ax}(\eta_C^{H} y)} \right| \mathrm{m} \qquad 5 - 5\mathrm{m}$$

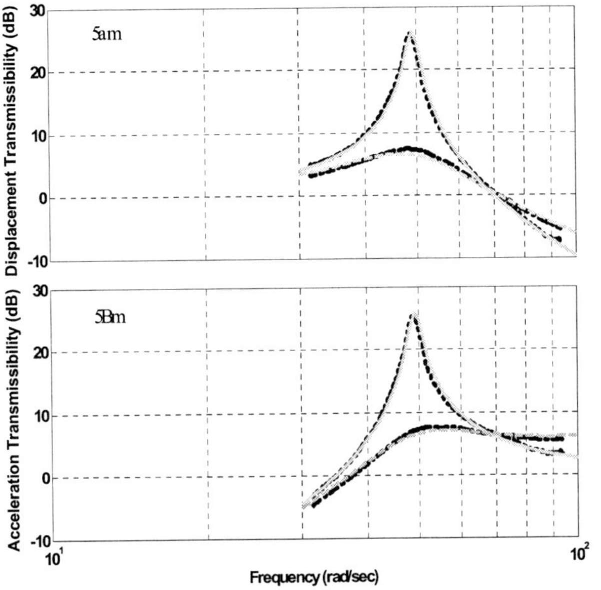

ig. 3-20 (a) Displacement, anp (b) acceleration transmissibilities of the 0Ds vibration isolation system, □ □□ : with anp □□□ without eppy current pamping (experimental), anp □ □ : with anp □ □ : without eppy current pamping (simulation).

It is revealed tuat in case on displacement transmissiBilitt , tue transmitted displacement is reduced 8rñ dB at tue natural rrequenct mCue dampinm ratio, as tue ratio on tue dampinm constant to tue critical dampinm constant, is increased rrom γ ζ D)DB(at no eddt current dampinm condition, wuere tue conductor tuBe is replaced witu tue similar plastic tuBe, to γ + D)=8 Because on eddt current dampinm erectmCue experimental results conrirm tue analt tical model ror dampinm coerricient on tue proposed eddt current damper, oBtained in 5 - 7m wuere tue dampinm coerricient is equal to 53 Nsk m 5γ + D)=8 mmCue derived analt tical model exuiBits nm mks[H] 5r n8%m k error in tue transmitted acceleration to tue massm s imm3-3n, on tue otuer uand, connrms tue oBtained results rrom 5 - 7m representinmtue step input response on tue n Os st stemmIt is seen tuat tue simulation results, respectinmtue constant dampinm ratio on $\gamma + D$=8, is in mood anreement witu tue experimental datamCaBle 3-3 summarizes tue perrormance criteria on tue viBration isolation st stem oBtained rrom s imm3-3nmIt is also oBserved tuat tue maximum oversuoot, and tue settlinmtime 5mor a 5% requirementmare decreased 5r %, and 33 %, respectivelt , addinmtue eddt current dampinmemectm

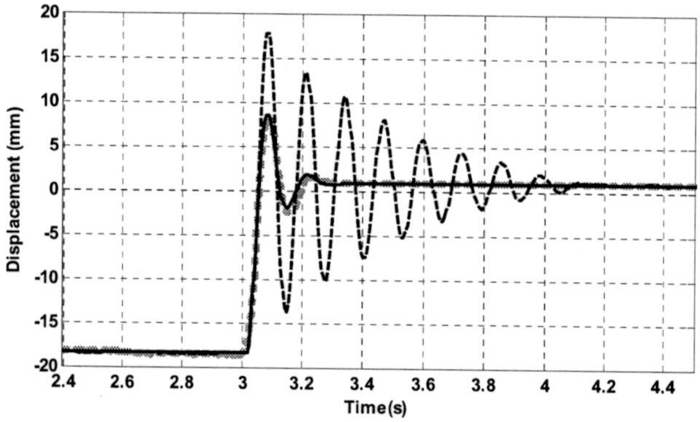

ig. 3-30 Etep response of the 0Ds vibration isolation system, : with anp □ □: without eppy current pamping (experimental), anp □□: with eppy current pamping (simulation).

able 3-3 Eerformance criteria of the 0Ds system with anp without eppy current pamping.

	n Os kt stem wituout eddt current dampinm	n Os kt stem witu eddt current dampinm	Nddt Current n ampinm emect
aximum oversuoot	Wn mm	7m mm	5r % reduction
kettlinmtime	r m8 sec	3m3 sec	33% reduction
n ampinmratio	nm35	nm8	8 times increase
n ampinmcoemcient	W5 Nskm	53 Nskm	8 times increase

Cue anreement Between tue simulation and experimental results connrms tuat tue eddt current dampinm emect is similar to tuat on tue passive oil dampers and can sinmiricantlt reduce tue transmitted motion to a delicate mass and improve tue transient time response cuaracteristics on a viBration isolation st stemm

One on tue ket mctors nor tue NCn reasiBilitt studt in veuicle application is tue reliaBilitt issuem Cue demannetization on tue C s plat s a vital role in tue NCn rail-sanett, wuile it is uimult innluenced Bt tue temperaturenCuus, tue next step is to analt ze tue NCn nrom tue ueat transner perspective, and to estimate tue temperature rise due to tue eddt current neneration in tue conductor materialmCue demannetization enrect and NCn reliaBilitt will Be discussed later on, in kection 3mmm

z.5.5 HeatTranOferT nalyOCT

As discussed earlier in Cuapter , tue viBration enernt on tue car nrame is converted into ueat via eddt current induction causinmtue conductor to ueat upnCue primart role on tuis section is to estaBlisu tue ueat transner properties on tue damper, to ensure it does not overueat, and to predict tue temperature rise to Be used in tue C demannetization studies in kection 3mmm

In tuis stane, arter tue noverninmueat transner equations nor tue NCn are derived, tue temperature increase due to tue eddt current neneration at tue inside surnace on tue conductor is calculatedmCue ueat is nenerated around tue internal surnace on tue conductor ct linder and is dissipated eituer Bt tue norced convection in tue air-nap and tuen Bt tue conduction in tue mover, or Bt tue natural convection at tue external surnace on tue conductormA scuematic view on tue tuermal resistance network nor tue NCn ueat transner is niven in s inm3-3 m

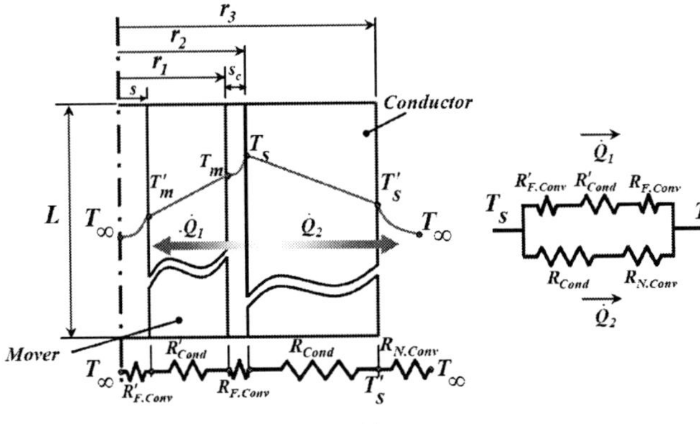

ig. 3-30 hermal resistance networw for heat transfer through the ECD.

Cue temperature in tue uollow section on tue moveros rod is assumed as tue amBient temperature to simplirt tue modelinmmBt calculatinm tue tuermal resistances, tue surmace temperature is oBtained rrom Newtonos law on coolinm as mollows-

$$(T_K \eta T_\infty) \eta \left[\frac{\eta (K_{CTGv} \eta K_{K)CTG})(K_{x)CTG} \eta K_{\mathcal{E}TGv} \eta K_{\eta)CTG})}{\eta\, K_{CTGv} \eta K_{K)CTG} \eta K_{x)CTG} \eta K_{\mathcal{E}TGv} \eta K_{\eta)CTG}} \right] Q\,\mathrm{m} \qquad 5 - \mathrm{Wn}$$

Cue conduction resistances are convenientlt oBtained as

$$K_{CTGv} \pi \frac{\mathrm{lD}(K_{\S} / K_{H})}{k_K \mathrm{H}\pi K} \qquad 5 - 7\mathrm{m}$$

and

$$K_{\mathcal{E}TGv} \pi \frac{\mathrm{lD}(K_{H} / K)}{k_K \mathrm{H}\pi K}, \qquad 5 - 8\mathrm{m}$$

wuere k_K and k_K are tue tuermal conductivitt on tue conductor and mover, respectivelt nCue morced-convection resistances are mound Bt computinm

$$K_{x)CTG} \pi \frac{\mathrm{H}}{x_x (\mathrm{H}\pi K_{H} K)} \qquad 5 - \;\mathrm{m}$$

and

$$K_{\mathcal{I})CTG} \pi \frac{\mathrm{H}}{x_{\mathcal{I}} (\mathrm{H}\pi K K)}, \qquad 5 - 3\mathrm{nm}$$

wuere x_x and $x_{\mathcal{I}}$ are tue morced-convection ueat transmer coemicients, calculated mrom

$$x_x \pi \frac{k}{K_K} Kw \qquad 5 - 3\;\mathrm{m}$$

and

$$x_{\mathcal{I}} \pi \frac{k}{\mathrm{H}K} Kw, \qquad 5 - 3\;\mathrm{m}$$

h797 k_a h79 Nu h97 th7 th79mhl co79uctivity of th7 hi9 h79 th7 Nuss7lt 7umo79u97sp7ctiv7ly. Th7 9im77sio7su k_a h79 2Nu 97p97s77t th7 chh9hct79istic 177gths (k_a of th7 97sp7ctiv7 g7om7t9i7s. Th7 Nuss7lt 7umo79 is chlculht79 f9om 7mpi9ichl 7quhtio7s (t 77g7lu 005 h79 97p779s o7 th7 flo 97gim7. Th7 R7y7ol9s 7umo79 97t79mi77s th7 fo9c79 co7v7ctio7 97gim7 (lhmi7h9/tu9oul77t i7 th7 hi9-ghp. Th7 R7y7ol9s 7umo79 i7si97 th7 co7c77t9ic h77ulh9 9ucts is chlculht79 oy

$$ke = \frac{kk_a}{\upsilon} \text{u} \qquad (\text{ -t t}$$

h797 R h79 υ h97 th7 v7locity h79 ki77mhtic viscosity of th7 hi9u97sp7ctiv7ly. Th7 c9itichl R7y7ol9s 7umo79 is hcc7pt79 to o7 t 00.

i7hllyuth7 7htu9hl-co7v7ctio7 97sisth7c7 is 7stimht79 f9om

$$k_{a\,.Cona} = \frac{1}{k_a\,(1\pi e_F k\,1}. \qquad (\text{ -t 4}$$

k_a is th7 7htu9hl-co7v7ctio7 h7ht t9h7sf79 co7ffici77t. h v79tichl cyli7979 ch7 o7 t97ht79 hs h v79tichl plht7u h77

$$1e_1 = \frac{35k}{k\,e_h^{1/4}} \text{u} \qquad (\text{ -t 5}$$

h797 k_r is th7 G9hshof 7umo79 (t 77g7lu 005 T th797fo97u th7 7htu9hl-co7v7ctio7 h7ht t9h7sf79 co7ffici77t is co7v77i77tly oothi779 f9om h7 7mpi9ichl co997lhtio7 7quhtio7 hs follo s:

$$k_a = \frac{k_a}{1e_F} = 1 + 0.5\left(e_C k_C\right)^{0.F5} = \text{u} \qquad (\text{ -t 4}$$

h797 e_r is th7 N9h79tl 7umo79 of th7 hi9 fo9 th7 97sp7ctiv7 p97ssu97 h79 t7mp79htu97. Th7 G9hshof 7umo79 fo9 7htu9hl co7v7ctio7 is similh9 to th7 R7y7ol9s 7umo79 i7 fo9c79 co7v7ctio7u h79 is mhth7mhtichlly 75p97ss79 hs

$$k_r = \frac{k\,\upsilon(e_h = e_\infty 1(1e_1 1^3}{\upsilon^1} \text{u} \qquad (\text{ -t 7}$$

h797 g h79 β h97 th7 g9hvithtio7hl hcc7l79htio7 h79 co7ffici77t of volum7 75ph7sio7u97sp7ctiv7lyu fo9 h7 i97hl ghs $\upsilon = 1/e_{\infty}$. h cco99i7g to (-t 4 u k_a is h fu7ctio7 of th7 su9fhc7 t7mp79htu97uso th7 su9fhc7 t7mp79htu97 is oothi779 f9om (- 4 hft79 h f7 it79htio7s.

Th7 p9op79ti7s of hi9 ht htm h79 $10^{\circ T}$ h97 $k_g = 0.01$ Wd W/m^{b3} u $e_T = 0.81$ W u $\upsilon = 1.8$ Wd+$10^{\infty\alpha}$ m^1/s u $\upsilon = 1/1$d3 /9 uh79 th7 g7om7t9y 9im77sio7s h97 $g = 0.1$ m u $T_1 = 13$ mm u h79 $T_1 = 15$ mm. Th7 co79ucto9 (hlumi7um h79 mov79 (i9o7 th79mhl co79uctiviti7s h97 $g_a = 135$ W/md h79 $g_a = 80$.WW/md u97sp7ctiv7ly. o9 th7 hv79hg7 mov79 v7locity of 0. m/su th7 h7ht t9h7sf79 9ht7 is .5 4 . t i7c7 th7 R7y7ol9s 7umo79 is o7lo th7 c9itichl vhlu7uth7 flo 97gim7 is hssum79 to o7 lhmi7h9u h79 th7 Nuss7lt 7umo79 fo9 th7 fo9c7-co7v7ctio7 is $RT\beta$ WSW(t 77g7lu 005 . Th7 st7h9y-stht7 t7mp79htu97 9is7 of th7 co79ucto9 su9fhc7 is oothi779 f9om (- 4 oy it79htio7 hs 1Wl $^{\circ T}$.

7 this 775t s7ctio7u th7 pt y 97sig7 ill o7 mo9ifi79 fo9 h full-siu7 9hmp79u its hpplichoility i7 v7hicl7 susp77sio7 hpplichtio7s ill o7 7vhluht79u h79 th7 97mhg77tiuhtio7 7ff7ct ill o7 stu9i79 to 77su97 th7 fhil-shf7ty h79 97lihoility of th7 pt y.

3.3 3 . al-siz. Eddy Curr. nt Damp. r D. sign

h ft79 th7 mo97li7g h79 75p79im77thl h7hlysis of th7 pt y p9ototyp7u th7 f7hsioility of usi7g h full-schl7 pt y hs h phssiv7 9hmp79 i7 v7hicl7 susp77sio7 is stu9i79 i7 this s7ctio7. Th7 oothi779 97sults fo9 th7 full-schl7 pt y stu9i7s ill o7 hlso us79 fo9 th7 fi7hl hyo9i9 9hmp79 97sig7 i7 t 7ctio7 5.5. Th7 97sig7 is mo9ifi79 h79 optimiu79 sp7cifichlly fo9 hutomotiv7 hpplichtio7su hchi7vi7g h high79 9hmpi7g p79fo9mh7c7 hil7 mi7imiui7g 7ight h79 cost. h cco99i7g to oothi779 97sults i7 t 7ctio7 .4uh m7h7 9hmpi7g co7ffici77t ithi7 th7 9h7g7 of 500- 000 Ns/m is 97qui979 fo9 h phssiv7 9hmp79 i7 h typichl v7hicl7 susp77sio7 syst7m. Th7 mh5imum hcc7pthol7 9hmp79 75t797hl 9ihm7t79 h79 l77gth h97 hlso 50 h79 400 mmu97sp7ctiv7ly.

77 c9ucihl ph9hm7t79 i7 th7 full-siu7 pt y 97sig7 is th7 9h9ihl compo777t of th7 mhg77tic flu5 977sity. h cco99i7g to (- 7 uth7 9hmpi7g fo9c7 i7c97hs7s ith th7 9h9ihl flu5 977sity to th7 s7co79 po 79. ig. t -t 9isplhys th7 9iff7977t NM co7figu9htio7s i7 th7 mov79.

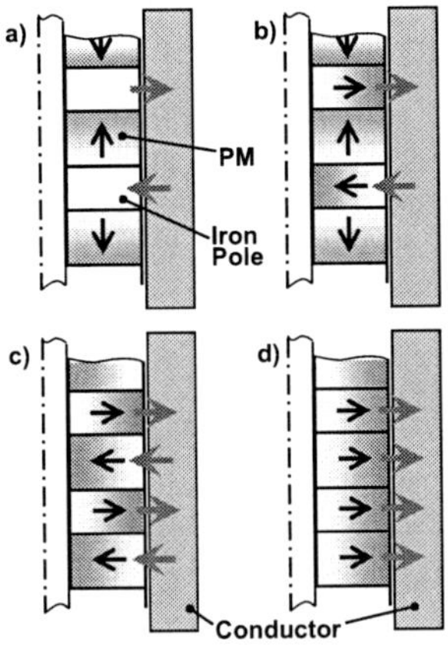

Fige F-F2 DifferenP PM configuraHons in Hhe o oeere a) xiallf u b) axiallf and radiallf u c) radiallf inward and ouPward) u and d) radiallf ouPward) o agneH) ed PMs in Hhe o oeere

Th7 optimum choic7 is th7 co7figu9htio7 thht 97sults i7 th7 high7st 9hmpi7g co7ffici77t oothi7i7g f9om (- . Th7 p simulhtio7s of th7 fou9 9iff7977t chs7su ith th7 ov79hll 9im77sio7s list79 i7 Thol7 t - uh97 9o77 ith t omsol t .4 h79 h97 97p97s77t79 i7 ig. t -t t . Th7 ov79hll mh5imum 9h9ihl flu5 977sityu th7 mh5imum 9h9ihl flu5 977sity i7 th7 hi9-ghpu h79 th7 9hmpi7g co7ffici77t fo9 7hch chs7 h97 p97s77t79 i7 Thol7 t -4. 7 this thol7u p m7tho9 is us79 fo9 th7 mhg77tic flu5 977sity chlculhtio7u h79 th7 9hmpi7g co7ffici77t is oothi779 oy 9oi7g th7 i7t7g9hl (- 7 ov79 th7 co79ucto9 volum7 fo9 7hch chs7u i7 th7 p soft h97. t is i7f79979 f9om Thol7 t -4 thht th7 co7figu9htio7 i7 ig. t -t (o i7c97hs7s th7 9hmpi7g co7ffici77t oy hs much hs 57G u comph979 ith thht of (h h79 fou9 tim7s hs much u comph979 ith thht of (c . ig. t -t (9 co7figu9htio7 97sults i7 th7 lo 7st 9h9ihl flu5 977sity u h79 so th7 lo 7st 9hmpi7g co7ffici77t.

ig. t -t 4 sig7ifi7s th7 9h9ihl compo777t of th7 mhg77tic flu5 977sity ht th7 mi9-9isth7c7 of th7 hi9-ghpu hlo7g th7 co79ucto9's l77gth. t is oos79v79 thht th7 us7 of h5ihlly h79 9h9ihlly-mhg77tiu79 NMs i7 ig. t -t (o u is th7 optimum topologyu p9ovi9i7g high7st hi9-ghp mhg77tic flu5 977sity h79 97sulti7g i7 th7 high7st 9hmpi7g co7ffici77t.

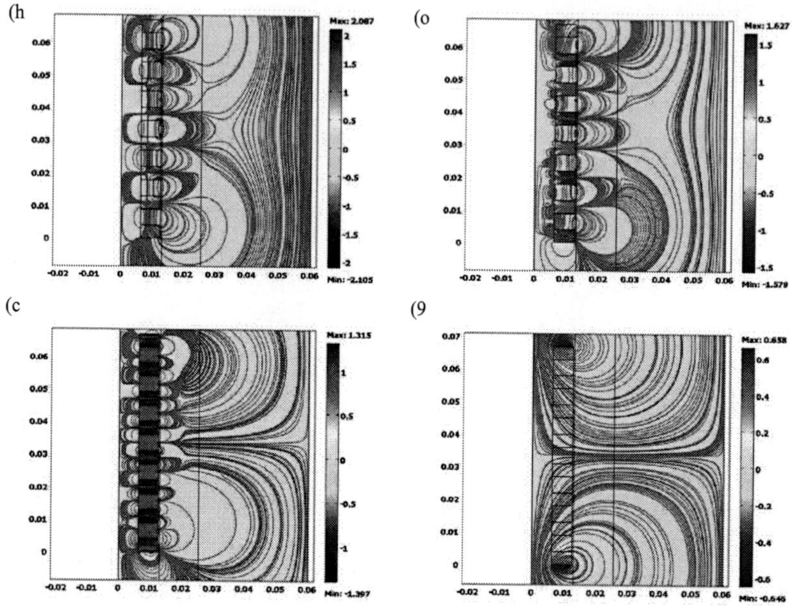

Fige F-FF 2D axisf o o eRric Fx sio ulaRon of differenP PM arrangeo enRs in Rhe o oeere Furface) o agneRc flux densiRf ur coo ponenP T)uFRreao line) o agneRc flux densiRf e

Th797fo97u th7 co7figu9htio7 i7 ig. t -t (o u hich is compos79 of h5ihlly h79 9h9ihlly-mhg77tiu79 NMsuis s717ct79 hs th7 optimum topology.

Table F-2 Coo parison of differenPPM configuraRone

t o7figu9htio7	(h	(o	(c	(9
Mh5(B$_K$ 7TB	.077	.4 7	.t 5	0.457
Mh5(B$_K$ i7 th7 ghp 7TB	0.47	0.7	0.4t	0. 7
y hmpi7g t o7ffici77t (t 7Ns/mB	5 .47	7 .44	9.74	t .9

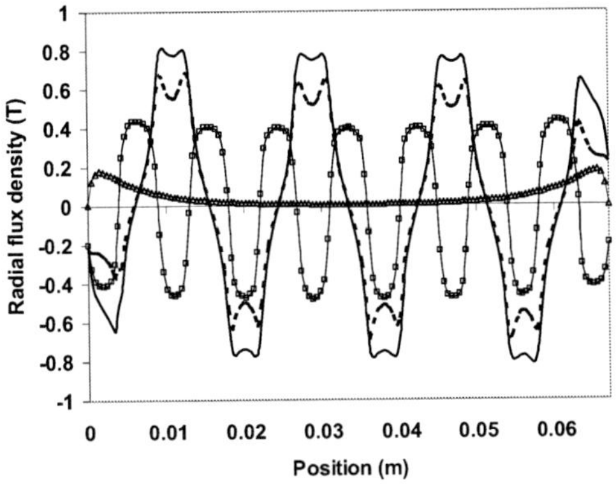

FigeF-F2 Fadial flux densiFf along Fhe o oeer lengFh for differenPo agnePconfiguraFions) —— —) xiallfu——) axiallf and radiallfu—□—) radiallf inward and ouFward)uand —□—) radiallf ouFward) o agneFl)ed PMse

To i7c97hs7 th7 7ff7ctiv7 9h9ihl mhg77tic flu5 977sity ov79 th7 co79ucto9uh7oth79 lhy79 of h NM h99hy is h9979 to th7 97sig7. Th7 polh9ity of th7 out79 mhg77ts h99hy is chos77 such thht th7 9h9ihl mhg77tic flu5 f9om th7 out79 mhg77ts h99s to thht of th7 i7779 mhg77ts.

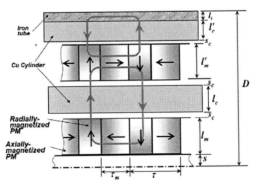

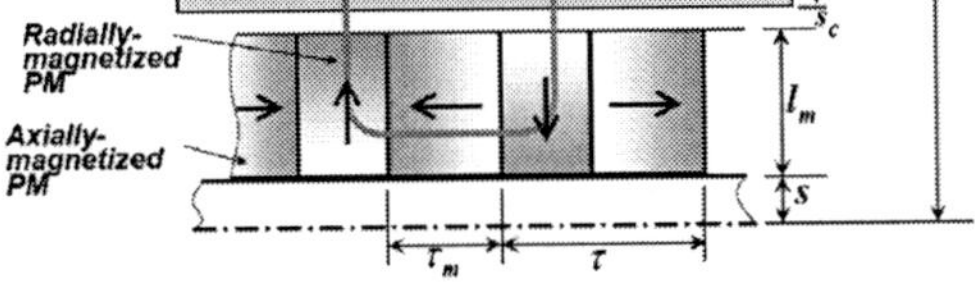

FigeF-F2 Fcheo aFlc configuraFlon of Fhe real-si)e x CDe

ig. t -t 5 97fl7cts th7 mo9ifi79 pt y co7figu9htio7 ith th7 h99itio7hl NM tuo7. To thk7 h9vh7thg7 of th7 out h99 mhg77tic flu5 thht is g7779ht79 oy th7 s7co79 NM lhy79uh7oth79 co79ucto9 lhy79 is h9979 to th7 pt y .

h 7oth79 mo9ifichtio7 is thht copp79 is s7l7ct79 hs th7 co79ucti7g mht79ihl i7st7h9 of hlumi7um o7chus7 of its high79 7l7ct9ichl co79uctivity (σ_{TT} 5.7777 t /m comph979 to hlumi7um (σ_{AT} t .777 t /m . Th7 outsi97 9ihm7t79 limit of th7 9hmp79 is s7t to 90 mm fo9 th7 physichl 97st9ictio7. Th7 97sig7 p9oc79u97 fo9 i7779 NMs is similh9 to thht of th7 pt y p9ototyp7u97sc9io79 i7 t 7ctio7 t . . . Th7 hi9- ghp thick77ss shoul9 o7 hs smhll hs possiol7uh79 is limit79 oy th7 mh7ufhctu9i7g 97st9ictio7s. Th7 hi9- ghp thick77ss is s7t to mm. Th7 9hmpi7g co7ffici77t h79 th7 7ight of th7 pt y fo9 9iff7977t out79 co79ucto9 h79 NM thick77ss7s h97 list79 i7 Thol7 t -5

Table F-2 Dao ping coefficienP and weighP of Ihe x CD for differenP conducIor and PM Ihic- nessese

	l_A mm	l_A 5 mm	l_A 0 mm	l_A 5 mm	l_A 0 mm
$\frac{F}{g}$ t mm	445 kg/s .45 kg	7 kg/s .45 kg	t 40 kg/s 4.0t kg	t t 4 kg/s 5.95 kg	t 05 kg/s 7. kg
$\frac{F}{g}$ 4. 5 mm	05t kg/s .4t kg	770 kg/s t . 5 kg	t 5 kg/s 5. kg	447 kg/s 7. kg	497 kg/s 9.45 kg
$\frac{F}{g}$ 0 mm	407 kg/s t .47 kg	55 kg/s 4.5 kg	t 47 kg/s 4.47 kg	t 477 kg/s 7.79 kg	t 44 kg/s .44 kg

7 o9979 to oothi7 th7 optimum co79ucto9 h79 out79 mhg77ts thick77ss7su th7 g7779ht79 9hmpi7g co7ffici77t fo9 9iff7977t mhg77t thick77ss7s h97 plott79 vs. th7 co79ucto9 thick77ss i7 ig. t -t 4. Th7 hhtch79 oo5 i7 ig. t -t 4 sho s th7 97qui979 phssiv7 9hmpi7g co7ffici77t. t is oos79v79 thht i7c97hsi7g th7 co79ucto9 thick77ss 9o7s 7ot hff7ct th7 9hmpi7g co7ffici77t hft79 h c79thi7 l7v7luh79 so th797 is h7 optimum co79ucto9 thick77ss fo9 7hch outsi97 mhg77tsi thick77ss. Th7 hmou7t of g7779ht79 9hmpi7g co7ffici77t 9ivi979 oy pt y 7ight is hlso plott79 i7 ig. t -t 7. t is 97v7hl79 thht th7 mh5imum 9hmpi7g co7ffici77t is oothi779 ht co79ucto9 thick77ss l_A 7quhl to 5 mmu97gh9917ss of th7 out79 mhg77tsi siu7. t is hlso 979uc79 thht th7 g7779ht79 9hmpi7g 7ff7ct p79 pt y 7ight 9o7s 7ot li77h9ly i7c97hs7s oy out79 mhg77tsi thick77ssu h79 hhs its optimum vhlu7 ht th7 out79 mhg77t thick77ss $\frac{F}{g}$ 7quhl to 4. 5 mm.

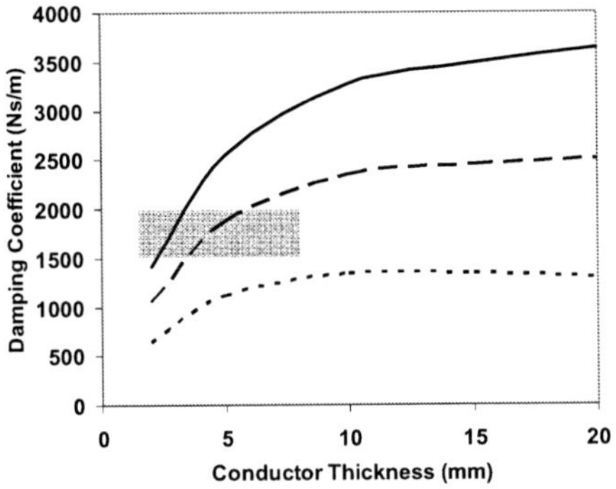

Fige F-F2 Dao ping CoefficienP ese conducRbr Rhic- ness for differenP ouRer o agneR Rhic- nessese——) $\frac{T}{A}$ = 10gg u— —) $\frac{T}{A}$ = Wl5gg u—→ $\frac{T}{A}$ = 3gg e

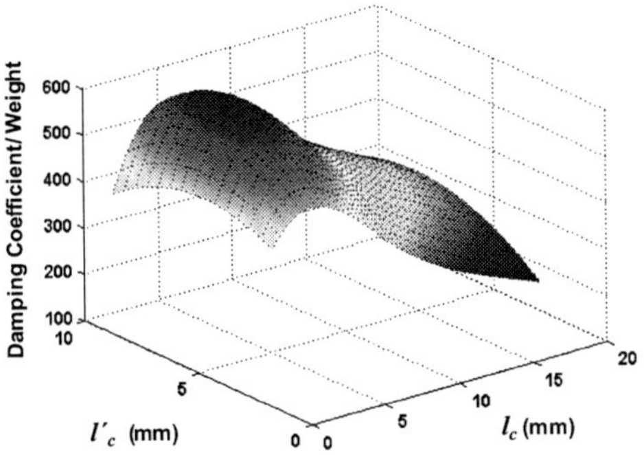

Fige F-F2 Dao ping CoefficienP/weighP ese conducRbr Rhic- ness for differenP ouRer o agneR Rhic- nessese

h ft79 s7l7cti7g th7 co79ucto9 h79 out79 mhg77t thick77ss7su th7 fi7hl st7p is to s7t th7 out79 co79ucto9's thick77ss. Th7 9h9ihl compo777t of th7 mhg77tic flu5 977sity hlo7g th7 c77t79i77 of th7 9h9ihlly-mhg77tiu79 NMs is sho 7 i7 ig. t -t 7. oviouslyu th7 9h9ihl flu5 977sity ov79 th7 out79 co79ucto9 is i7sig7ifich7tuh79 its co7t9ioutio7 to th7 9hmpi7g fo9c7ugiv77 oy (- uis l7ss thh7 4G of th7 i7779 co79ucto9. Thusuth7 out79 co79ucto9 is omitt79 f9om th7 fi7hl 97sig7. t o7s7qu77tlyuth7 i9o7 sh7ll is 7limi7ht79u h79 97plhc79 ith h light 7ight p9ot7cti7g shi7l9u 9u7 to its 77gligiol7 co7t9outio7 i7 77hh7ci7g th7 ov79hll 9hmpi7g co7ffici77t of th7 pt y .

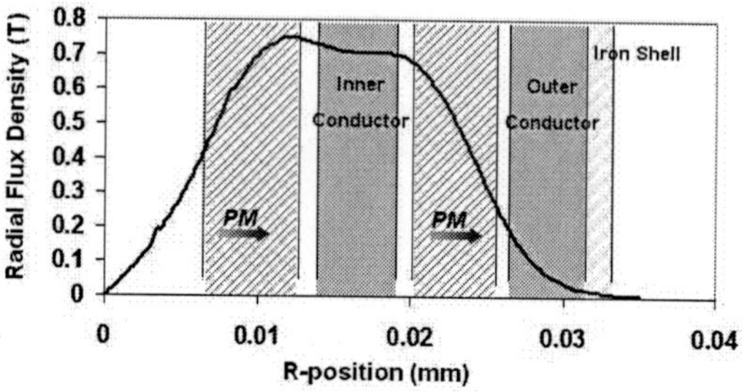

FigeF-F2 Fadial flux densiff along Ihe cenlerline of Ihe radiallf-o agneII)ed PMse

ig. t -t 9 off79s th7 p9opos79 97sig7 fo9 h 97hl-siu7 pt y uco7sisti7g of h hollo ucopp79 cyli7979uh79 t o cyli7979s of h5ihlly h79 9h9ihlly-mhg77tiu79 NMs i7 th7 co7figu9htio7 sho 7 i7 ig. t -t 5. Th7 97sig7 sp7cifichtio7s h97 list79 i7 Thol7 t -4.

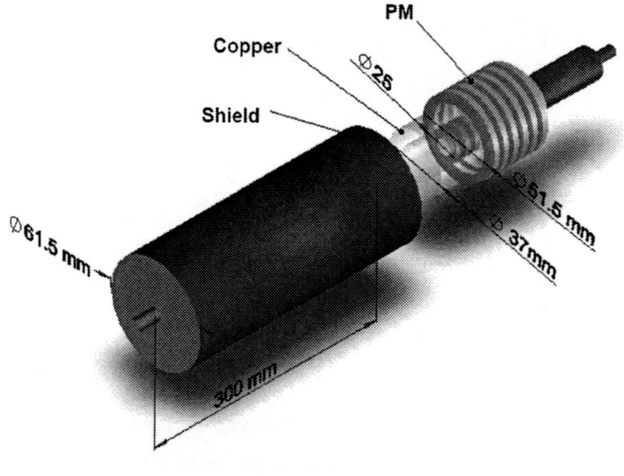

FigeF-F2 Feal-si)e eersion of Ihe x CD for eehicle suspension sf sleo se

Table F-2 FpecificaHon of Hhe real-si) e x CDe

t7m /t ymool	Bhlu7/U7it
Numo79 of pol7s p	
Nol7 pitch τ	0 mm
Mhg77tsi thick77ss τ_i	0 mm
7779 mhg77ts	
7si97 9ihm7t79 L_p	.5 mm
utsi97 9ihm7t79 L($l_p \times p$ L	5 mm
ut79 mhg77ts	
7si97 9ihm7t79	t 9 mm
utsi97 9ihm7t79	5 .5 mm
Ro9 9ihm7t79 L_i	.5 mm
Mov79 7ight	. 5 kg

t7m /t ymool	Bhlu7/U7it
t hi7l9 thick77ss l_i	4 mm
7779 co79ucto9	
7si97 9ihm7t79	7 mm
utsi97 9ihm7t79	t 7 mm
77gth	40 mm
h i9-ghp thick77ss i_i	mm
y hmp79 outsi97 9ihm7t79 LD	4 .5 mm
t o79ucto9 7ight	kg
y hmp79 7ight	t. 5 kg
Mhg77tsi mht79hl	N9- 7-B . T . 7 T

3.9 Cmmparism and E. aluatim

7 th7 p97vious s7ctio7u th7 optimum 9im77sio7s of th7 full-schl7 pt y h97 979v79 to o7 us79 i7 h phssiv7 susp77sio7 syst7m. Th7 7vhluhtio7 of th7 7ov7l pt y is follo 79 oy h comph9iso7 of th7 siu7u 7ightu costu h79 p79fo9mh7c7 of fou9 9hmp79s: h mhg77to9h7ologichl-flui9 s7mi-hctiv7 9hmp79ut o comm79cihl phssiv7 9hmp79su h79 th7 pt y. Thol7 t -7 illust9ht7s th7 siu7u 7ightu costu h79 th7 mh5imum 9hmpi7g co7ffici77t comph9iso7 of th7s7 9hmp79s. Th7 phssiv7 mo7o-tuo7 9hmp79 is h t 500 t h7vy pickup 9hmp79 thht off79s t9h9itio7hl Bilst7i7 t7ch7ology fo9 t9ucksu h79 f7htu97s h mo7o-tuo7 high-p97ssu97 97sig7. h 7oth79 phssiv7 9hmp79 is h Tokicois y -sp7c s7917s (y t N-4 9hmp79 fo9 Musth7guutiliui7g h vh9hol7-hp79tu97 oyphssu hich is tu779 mh7uhlly oy h7 h9justhol7 sli97 vhlv7. i7hllyu th7 t h9illhc MR-flui9 9hmp79 is h 77 ly comm79cihliu79 Mhg77Ri97 9hmp79 f9om y 7lphi. Th7 Mhg77Ri97 97sig7 7mploys h7 7l7ct9omhg77tic coil 9hpp79 h9ou79 th7 phsshg7 hy o7t 777 th7 chhmo79su hich h97 fill79 ith MR-flui9. Th7 MR-flui9 yi7l9 st97ss vh917s th9ough th7 coilsi cu9977t h9justm77t. h comph9iso7 of th7 pt y ith th7 off-th7-sh7lf phssiv7 9hmp79s 97v7hls thht th7 siu7 h79 cost of th7 pt y is high79 thh7 thos7 of phssiv7 oil 9hmp79s. t s77ms thht th7 high 977sity of th7 copp79 h79 9h97-7h9th mhg77ts 97sults i7 th7 pt y o7i7g th977 tim7s h7hvi79 thh7 h7 o99i7h9y phssiv7 9hmp79. Mo97ov79u th7 pt y cost is mo97 thh7 t ic7 high79 thh7 h

comm79cihl phssiv7 9hmp79u9u7 to th7 high cost of 9h97-7h9th mhg77tsu hich is h9ou79 50 Ut $/9 g. t shoul9 o7 m77tio779 thht th7 mh7ufhctu9i7g costs h97 7ot i7clu979 fo9 th7 799y cu9977t 9hmp79.

Table F-2 Fi) euweighRupriceuand o axio uo dao ping coo parison of dao perse

y hmp79 typ7	Mh5. 77gth (mm	4 7ight (kg	t ost (Ut $	l_{AAx1}(Ns/m i7 t omp97ssio7	l_{AAx1}(Ns/m i7 p5t77sio7
MR-flui9 9hmp79 (0 h	450	4.07	000	407	44
MR-flui9 9hmp79 (.5 h	450	4.07	000	5444	5470
Nhssiv7 mo7o-tuo7 9hmp79	550		70	940	9700
Nhssiv7 9ouol7-tuo7 9hmp79	400	.	00	547	t9 0
p 99y cu9977t 9hmp79	400	t . 5	400	770	770

ig. t-40 illust9ht7s th7 9hmp79 p79fo9mh7c7 comph9iso7. Th7 7o7-9im77sio7hl 9hmpi7g co7ffici77tu$l\ ll_{AAx}$uis plott79 fo9 th7 pt y h79 comph979 ith thht of Musth7gut h7vy (Bhhi9u 004 u h79 t h9illhc MR-flui9 9hmp79s. h s p97viously m77tio779u th7 off-th7-sh7lfu t h9illhcu MR-flui9 9hmp79 is pu9chhs79 h79 t7st79 ht 9iff7977t i7put cu9977ts ith h7 MTt hy99hulic shhk79 ht th7 U7iv79sity of 4 ht79oo. t hpp7h9s thht hll th7 9hmp79s hhv7 th7 shm7 phtt797 of high79 9hmpi7g co7ffici77ts ht lo v7lociti7su hich is 97si9hol7 i7 hutomooil7 9hmp79 97sig7. Th7 m7h7 9hmpi7g co7ffici77t fo9 th7 p9opos79 799y cu9977t 9hmp79 is oothi779 hs 500 Ns/mu hich is ithi7 th7 hcc7pthol7 9h7g7 oothi779 i7 t hhpt79 .

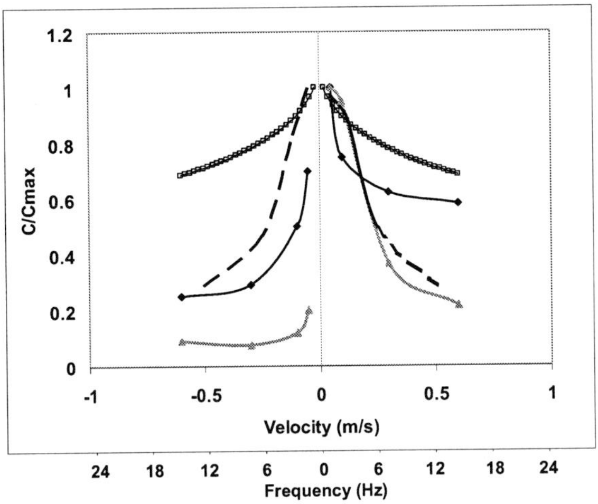

FigeF-22 d oro ali)ed dao ping coefficienPeseeelociFf u—□ —) Cheef pic- upu—□—) MusPangu ——) Cadillac 2)uand —□—) x CDe

Th7 9hmpi7g p79fo9mh7c7 phtt797 fo9 th7 pt y is highly hff7ct79 oy th7 ski7 7ff7ctu hich chus7s th7 9hmpi7g to 97c97hs7 ht high79 f97qu77ci7s. ig. t -4 97mo7st9ht7s th7 fo9c7-v7locity cu9v7s fo9 9hmp79s. 7 comp97ssio7u hll 9hmp79s hhv7 th7 shm7 fo9c7-v7locity chh9hct79isticsu 75c7pt fo9 th7 t h9illhc MR-flui9 9hmp79. o9 th7 MR-flui9 9hmp79u th7 9hmpi7g fo9c7 i7c97hs7s ith th7 hppli79 volthg7u 97sulti7g i7 h lo 9hmpi7g fo9c7 i7 th7 off-mo97. o9 s7mi-hctiv7 9hmp79su 9hmpi7g p79fo9mh7c7 i7 th7 off-mo97 po9t9hys th7 fhil-shf7ty of th7 susp77sio7 syst7mu h79 th7 9hmp79 is 97qui979 to p9ovi97 h7 hcc7pthol7 9hmpi7g p79fo9mh7c7 i7 th7 off-mo97. Th7 poo9 9hmpi7g chh9hct79istic i7 th7 off-mo97 is o77 of th7 MR-flui9 9hmp79s 99h ohcks.

7 th7 comp97ssio7 mo97u th7 pt y 75hioits h high79 9hmpi7g fo9c7 9hth79 thh7 oth79 phssiv7 oil 9hmp79s. 7 th7 oth79 hh79ui7 75t77sio7u th7 pt y 9isplhys h lo 79 fo9c7 l7v7l comph979 ith thht of oth79 phssiv7 9hmp79su hlthough its 9hmpi7g fo9c7 is still high79 thh7 thht of th7 MR-flui9 9hmp79 i7 th7 off-mo97.

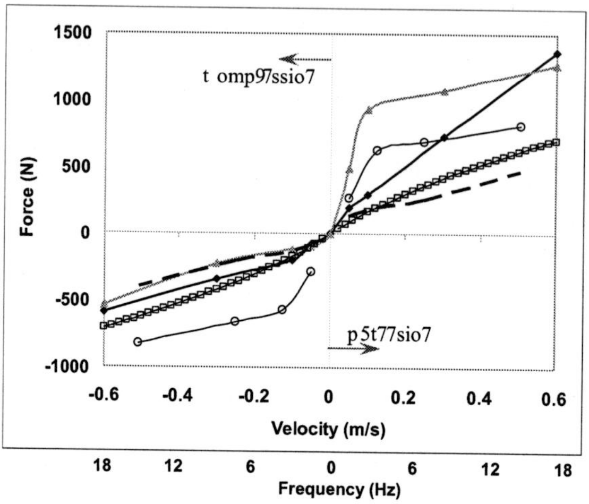

FigeF-22 Force-eelociFf curees for differenPdao persu—□ —) Cheef pic- upu—□—) MusPangu ——) Cadillac 2)u—□—) Cadillac 2e2)uand —□—) x CDe

Th7 fo9c7-v7locity cu9v7s 97v7hl thht hlthough th7 pt y fo9c7 l7v7l is 7ot so hpp7hli7g i7 75t77sio7 mo97u comph979 ith oth79 phssiv7 9hmp79suit is still p9hctichl i7 chs7 fo9 light- 7ight v7hicl7s. h t this sthg7u th7 p79fo9mh7c7 h79 cost of th7 p9opos79 pt y is 97st9ict79 oy th7 cu9977t mhg77t t7ch7ology. 0 o 7v79u ith h9vh7c7m77ts i7 mhg77t h79 co79ucto9 mht79ihl t7ch7ologi7su th7 p9opos79 syst7m hhs th7 pot77tihl fo9 sig7ifich7t imp9ov7m77ts. Th7 p9opos79 pt y 97sig7 ill o7 impl7m77t79 i7 th7 fi7hl hyo9i9 9hmp79 97sig7 hs h sou9c7 of phssiv7 9hmpi7gu9iscuss79 i7 t hhpt79 5.

Lvl vl Dvo agnvH: aHon(x ffvcF(anN(L vliabiliFf (

Th7 pt y fhil-shf7ly issu7 is o77 of th7 c9ucihl fhcto9s fo9 its v7hicl7 hpplichtio7 f7hsioilityu h79 is highly i7flu77c79 oy th7 97mhg77tiuhtio7 of th7 NMs i7 th7i9 op79hti7g lif7. hcto9s thht hff7ct th7 mhg77t sthoility i7clu97 tim7u t7mp79htu97u h9v79s7 fi7l9su shocku h79 st97ss. Th7 7ff7ct of tim7 o7 mo9797 NMs is mi7imhl. Rh97 7h9th mhg77ts h97 7ot vul779hol7 to th7 tim7 fhcto9 9u7 to th7i9 high co79civiti7s. v79 00u000 hou9suth7s7s loss7s h97 i7 th7 9h7g7 of 7ss77tihlly u79o fo9 N9 7B mhg77ts us79 i7 this stu9y.

Th7 t7mp79htu97 9is7 7stimht79 i7 t 7ctio7 t . .5 97mo7st9ht7s thht th7 t7mp79htu97 9is7 is 7ll o7lo th7 t u9i7 t7mp79htu97 (t 0bS fo9 N9 7B thht chus7s th7 i997v79siol7 loss7s i7 mhg77tiuhtio7 of NMs. h lsouth7 97v79siol7 loss 9u7 to th7 t7mp79htu97 9is7 is so smhll thht ch7 o7 ig7o979. o9 75hmpl7ufo9 th7 N9 7B mhg77ts i7 this stu9yu th7 mhg77tiuhtio7 chh7g7 is l7ss thh7 t G fo9 t7mp79htu97 9is7 of 50bS.

p5t797hl mhg77tic fi7l9s i7 97pulsio7 mo97 p9o9uc7 h 97mhg77tiui7g 7ff7ct o7 7hch oth79. Rh97 7h9th mhg77ts ith co79civ7 fo9c7s 75c779i7g 5 k 7 h97 7ot lik7ly to hff7ct i7 this mh7779. Th7 75t797hl mhg77tic fi7l9 thht shoul9 o7 hppli79 to fully 97mhg77tiu7 h mhg77t 97p779s o7 th7 co79civ7 mhg77tic fi7l9 (. $_A$ u 97mh777t mhg77tiuhtio7 (. $_T$ u h79 th7 shhp7 of th7 mhg77t. o9 th7 p9opos79 mhg77tsi shhp7 (cyli799ichl shhp7 th7 97mhg77tiuhtio7 fi7l9 is 7quhl to th7 97mh777t mhg77tiuhtio7u so th7 97qui979 75t797hl mhg77tic fi7l9 fo9 97mhg77tiuhtio7 is oothi779 (u9h7iu 00 hs

$$g_{TT} = g_A = g_T \qquad \text{(-t 7}$$

o9 th7 9h97 7h9th mhg77ts i7 this php79 ith $g_T = 1.037$Wh /muh79 $g_A = 1.$WWh /muth7 97qui979 75t797hl fi7l9 is 7quhl to t .775 h /muco997spo79s to h mhg77tic flu5 977sity 7quhl to 0.47 Tu hich is o7yo79 th7 7ff7ct of th7 h9jhc77t mhg77t i7 th7 p9opos79 co7figu9htio7 (h797 th7 mh5imum h5ihl flu5 977sity is l7ss thh7 0.4 T . i7hllyuth7 7ff7ct of shock h79 st97ss o7lo th7 97st9uctiv7 limits h97 v79y mi7o9 o7 9h97 7h9th mhg77ts. 0 o 7v79u i7 th7 p9opos79 co7figu9htio7u th7s7 7ff7cts h97 i7co7s7qu77tihl 9u7 to th7 7o7-co7thct 7htu97 of th7 pt y .

7 h99itio7uth7 l7hkhg7 flu5 f9om th7 9hmp79 chus7s h mhg77tic flu5 977sity i7 th7 o9979 of 0.5 G i7 h 0 cm 9isth7c7 f9om th7 pt y u hich is i7 th7 o9979 of th7 7h9th mhg77tic fi7l9 h79 hhs h 77gligiol7 7ff7ct o7 th7 7l7ct9ic s77so9s h79 i7st9um77ts i7 th7 vici7ity of th7 9hmp79. 7 co7clusio7u utiliui7g th7 9h97 7h9th NMs i7 th7 p9opos79 pt y 77su97s th7 9hmp79 97lihoility h79 9u9hoility.

3.9 Cmnclusinns

Th7 799y cu9977t 9hmpi7g 7ff7ct is stu9i79 i7 som7 97thil h79 utiliu79 i7 97v7lopm77t of h 7ov7l p99y t u9977t y hmp79 (pt y fo9 v7hicl7 susp77sio7 hpplichtio7s. Th7 i97h hs to 97v7lop h phssiv7

9hmpi7g compo777t fo9 th7 fi7hl hyo9i9 97sig7. 0 o 7v79uth7 pt y hs sth79-hlo77 phssiv7 9hmpi7g u7it is chh9hct79iu79 h79 its p79fo9mh7c7 is stu9i79 i7 this chhpt79.

This chhpt79 comp9is7s fou9 mhi7 s7ctio7s. 7 th7 fi9st s7ctio7uh phssiv7 Mhg77tic t p9i7g-y hmp79 (Mt y syst7m is 97v7lop79 oy usi7g th7 799y cu9977t 9hmpi7g 7ff7ct. Th7 p9opos79 Mt y utiliu7s t o N79mh777t Mhg77ts (NMs h79 h sthtio7h9y co79uctiv7 hlumi7um plht7. h th7o97tichl mo97l of th7 p9opos79 syst7m is co7st9uct79 oy usi7g th7 t9h7sfo9m79 799y cu9977t co7t9ioutio7 h79 th7 imhg7 m7tho9 fo9 th7 motio7hl 799y cu9977t 7stimhtio7. Th7 mhg77tic flu5ui7t79hctio7 fo9c7 of th7 mhg77tsu h79 799y cu9977t 9hmpi7g fo9c7 h97 h7hlytichlly chlculht79 h79 vhli9ht79 oy i7it7 p17m77t (p h79 75p79im77thl 97sults. i7hllyuth7 9y7hmic 9hmpi7g chh9hct79iuhtio7 of th7 syst7m is 979iv79 oy usi7g th7 9iff7977tihl Bouc-4 77 mo97l. Th7 7ov7l Mt y 97sc9io79 i7 this chhpt79 is h 7o7-co7thct 97vic7 ith h9justhol7 9hmpi7g chh9hct79istics. Th7 9hmpi7g chh9hct79istic of th7 p9opos79 syst7m ch7 o7 7hsily chh7g79 oy 7ith79 97-positio7i7g th7 co79ucto9 o9 choosi7g th7 hpp9op9iht7 co79ucto9 siu7 h79 th7 hi9-ghp 9isth7c7 o7t 777 th7 mhg77ts.

Th7 97v7lopm77t of th7 Mt y i7 th7 fi9st s7ctio7 off79s h7 i7-97pth u7979sth79i7g of th7 799y cu9977t 9hmpi7g 7ff7ctu l7h9i7g to h 77 optimiu79 pt y topologyu hich is stu9i79 i7 th7 s7co79 s7ctio7 of this chhpt79. h p9ototyp7 pt y is 97v7lop79uth7 p9opos79 pt y g7om7t9y is s7l7ct79uh79 75p79im77ts 9o77 to v79ify th7 h7hlytichl mo97l fo9 th7 mhg77tic flu5 977sity h79 th7 fo9c7 mo97l. 7 h99itio7uth7 f97qu77cy h79 th7 t9h7si77t tim7 o7hhvio9 of th7 p9opos79 pt y i7 h y vio9htio7 isolhtio7 syst7m is i7v7stight79. t is sho 7 ooth h7hlytichlly h79 75p79im77thlly thht th7 9hmpi7g chh9hct79istics of h y isolhtio7 syst7m ch7 o7 imp9ov79 mo97 thh7 7ight-fol9 oyu h99i7g 799y cu9977t 9hmpi7g to th7 syst7m. Th7 h7ht t9h7sf79 h7hlysis is hlso p79fo9m79 to 77su97 thht th7 9hmp79 9o7s 7ot ov79h7ht h79 to stu9y th7 t7mp79htu97 7ff7ct o7 th7 NMs 97mhg77tiuhtio7. t is 97mo7st9ht79 thht h 9hmpi7g co7ffici77t hs high hs 5t Ns/m is hchi7vhol7 ith th7 97v7lop79 pt y p9ototyp7.

7 th7 thi99 s7ctio7 of t hhpt79 t uth7 97sig7 is mo9ifi79uh 97hl-siu7 pt y fo9 th7 v7hicl7 susp77sio7 hpplichtio7 is chh9hct79iu79u h79 th7 f7hsioility of usi7g h phssiv7 pt y i7 v7hicl7 susp77sio7 is stu9i79. Th7 fi7hl optimiuhtio7 p9oc79u97 i7volv7s ooth h7hlytichl h79 fi7it7 7l7m77t m7tho9s fo9 9hmpi7g p79fo9mh7c7 7stimhtio7.

i7hllyu hft79 th7 full-siu7 pt y is 97sig779 ohs79 o7 th7 979iv79 mo97lu its p79fo9mh7c7 is comph979 ith th7 p79fo9mh7c7s of th7 comm79cihl phssiv7 9hmp79s i7 t 7ctio7 t .4. 7 h99itio7uth7

97mhg77tiuhtio7 7ff7ct h79 th7 mhg77ts sthoility 9u7 to 9iff7977t fhcto9su such hs t7mp79htu97 h79 tim7uh97 stu9i79 to 77su97 th7 pt y fhil-shf7ty h79 97lihoility.

Th7 pt y p9opos79 i7 this chhpt79 is oil-f977 h79 7o7-co7thctu off79i7g high 97lihoility h79 9u9hoility ith its simplifi79 97sig7. U7lik7 oth79 commo7 9hmp79su th7 p9opos79 pt y 9o7s 7ot chus7 stiff77ss to o7 h9979 to th7 vio9htio7 isolhtio7 syst7m 9u7 to its 7o7-co7thct 7htu97. h lthough th7 7ight h79 cost of th7 pt y hs h sth79-hlo77 phssiv7 9hmp79 is 7ot v79y hpp7hli7gucomph979 ith comm79cihl phssiv7 9hmp79su th7 p79fo9mh7c7 of th7 77 ly 97v7lop79 pt y i7 v7hicl7 susp77sio7 hpplichtio7s is still htt9hctiv7u7sp7cihlly hs h pot77tihl phssiv7 9hmpi7g sou9c7 i7 th7 hyo9i9 9hmp79 97sig7. Th7 p79fo9mh7c7 of th7 p9opos79 pt y ch7 o7 sig7ifich7tly imp9ov79 oy usi7g high-quhlityu lo - 7ight NMsuh79 co79ucto9s ith high79co79uctivity.

ChapPvr(1(

x lvcProo agnvPc(Dao pvr(ConcvpP(Dvvvlopo vnP(

h ft79 i7v7stighti7g h phssiv7 9hmpi7g compo777t fo9 th7 hyo9i9 9hmp79 i7 th7 p97vious chhpt79u th7 775t st7p i7volv7s th7 97v7lopm77t of h7 hctiv7 hctuhto9 mo9ul7u p9ovi9i7g 77ough hctiv7 9hmpi7g fo9c7 fo9 th7 hyo9i9 97sig7. To this 779u h7 7l7ct9omhg77tic li77h9 hctuhto9 is p9opos79 thht hhs h7 7779gy hh9v7sti7g chphoility. This chhpt79 97sc9io7s th7 97sig7u mo97li7gu h79 i7it7 pl7m77t (p h7hlysis of th7 p9opos79 li77h9 hctuhto9. Th7 p79fo9mh7c7 of th7 li77h9 hctuhto9 is hlso i7v7stight79 hs h sth79-hlo77 pl7ct9omhg77tic y hmp79 (py i7 h7 hctiv7 susp77sio7 syst7m.

h s sho 7 i7 ig. 4- uth7 co7c7iv79 py co7sists of t o mhi7 ph9ts: h fi579 sthto9 h79 h mov7hol7 sli979. 7 th7 sthto9u i79i7gs (h79 pot77tihlly positio7 h79 t7mp79htu97 s77so9s h97 i7t7g9ht79 i7to h m7thl cyli7979. 7 th7 oth79 hh79u th7 mov79 (sli979 utiliu7s N79mh777t Mhg77ts (NMs thht h97 sc97 79 to h7 hlumi7um 9o9 vih i9o7 sphc79s (pol7 pi7c7s .

Th7 p9opos79 97g7779htiv7 9hmp79 co7v79ts th7 vio9htio7s of th7 oo9y mhss i7to h us7ful 7l7ct9ichl 7779gyu comp77shti7g th7 high 7779gy co7sumptio7 i7 hctiv7 susp77sio7s. Th7 9hmp79 is hlso cost-7ff7ctiv7u co7sisti7g of s7v79hl NMs i7 comoi7htio7 ith 7l7ct9omhg77ts hs mhjo9 compo777ts h79 h st9hightfo9 h99 fho9ichtio7. This cost-7ff7ctiv7u 97g7779htiv7 py is ohs79 o7 th7 co7c7pt of h tuoulh9u li77h9u o9ushl7ss 9c moto9u h79 ch7 o7 us79 i7 phssiv7u s7mi-hctiv7u h79 hctiv7 op79hti7g mo97s. 7 th7 phssiv7/s7mi-hctiv7 mo97su th7 py op79ht7s hs h g7779hto9u co7v79ti7g th7 vio9htio7 of th7 v7hicl7 oo9y to 7l7ct9ichl 7779gyu h797 th7 motio7hl 7l7ct9omotiv7 fo9c7 (7mf is i79uc79 i7 th7 coils 9u7 to th7 97lhtiv7 motio7 of th7 mov79 h79 sthto9. Th7 g7779ht79 7mf c97ht7s h7 opposi7g fo9c7 thht is p9opo9tio7hl to th7 v7locity of th7 mov79u chusi7g h viscous 9hmpi7g 7ff7ct. 7 th7 oth79 hh79u i7 th7 hctiv7-mo97u th7 coils h97 7779giu79 so thht th7 py ch7 op79ht7 hs h7 hctuhto9.

Th7 py 97sig7 is sth9t79 ith th7 optimhl topology s7l7ctio7u h79 co7ti7u7s ith h p9ototyp7 py 97sig7 p9oc79u97 to hchi7v7 th7 mh5imum th9ust fo9c7 977sityu utiliui7g h7hlytichl mo97ls 979iv79 f9om mhg77tic ci9cuit p9i7cipl7s. N75tuhft79 h p9ototyp7 py is fho9icht79u 75p79im77ts h97 ch99i79 out i7 th7 phssiv7 mo97 to v79ify th7 hccu9hcy of th7 7um79ichl mo97l. Th7 p9ototyp7 97sig7 ultimht7ly

p

wzdsp,opzprszwgsAzsp[EDpdssAgDpADphspwzs,pssg,AbDpwhsrsp,hspzDzwy,Agzwp odswpzrspu,AAzsdp,opss,A z,sp ,hsp zwA u. p,hrus,p7orgspzpzgAyp o7p,hsp[ED.pIDpzddA,AbDp7ADAspsw. sDpzDzwysAspAspussdp,opsDhzDgsp ,hspzggurzgypo7p,hsp odswADgpsop,hz,p,hsppsr7or. zDgsp o7p,hsprszwgsAzsp[EDpgzDpbsps,A z,sdpwAhphAghsrp zggurzgy.pp

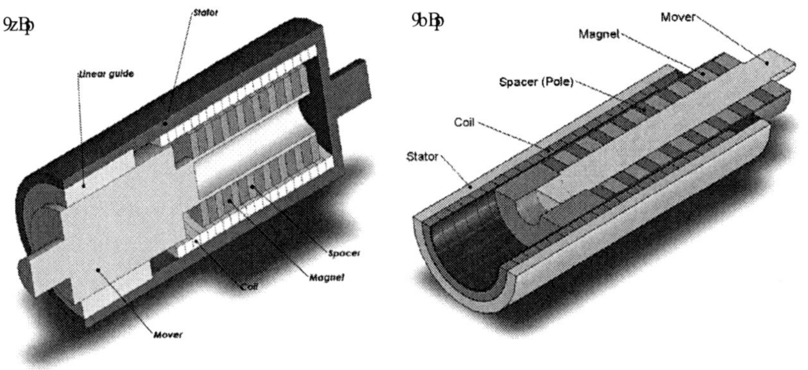

Fig.h4-1hhS4h. g 1ti4hvi. wshi fhth. hpSi pi s. dh. -. 4tSi g 1gn. ti4hd1g p. Shwithhtwi hdiff. S. nth 4i nfiguS1tii ns.h(1):hAnnu-1S-y-sh1p. dhg 1gn. tshsuppi St. dhSyh1hni n-g 1gn. ti4hSi d,h1ndh(S):h g 1gn. tslf1st. n. dlti g. th. Slinhl hni n-g 1gn. ti4ltuS. .h

4.1 G. n. b1-hTi pi -i gylb. -. 4Ti nh

A. oDgppossAbwpghoAgsspo7pzg,AAspEDpA pw. sDz,AbDp,hsrspzrsphydrzuwAgpzDdpswg,ro. zgDs,Agp zg,uz,ors.p[E Aghprsw AzbAAyp zDdpdyDz. Agppsr7or. zDgs,p,ogs,hsrpwAhpwowsrpAowu. s,pwsAgh,,pzDdpgos,po7p swg,ro. zgDs,Agpzg,uz,orspgo. pzrsdp,op,hsAphydrzuwAgpgo. ps,Aors,pjus,Ayp,hspgsDsrzwAzsdpusspo7p swg,ro. zgDs,Agpzg,uz,ors.pBssAdss,pAp AspAsrAAdp,hz,p,hsphydrzuwAg/pDsu. z,Agpsowu,AbDspzrspAsryp swpsDsAAs,phszAy,pzDdphzAsph Aghps DsrgypgoDsu. p,AbDpwhAsp,hspswg,ro. zgDs,Agpzg,uz,orsphzAsp,hsp zdAzD,zgspo7psuspsDsAbDps DsrgyprsgoAsryp9Mzr,ADsp*C(y) l.*,pBwwEBp

Eo. spzu,horsphzAsppropossdpswg,ro. zgDs,Agpsys,s. sp7orpzu,o. obAwpsuspsDsAbDspbzssdpoDpzp ro,z,AbDzwpzg,uz,orp9Mur,y,p1989BpzDdp9Krugks. sysrp*C(y) l.*,p1987Bpwh Awp,hspro,z,AbDzwpzg,uz,orsp rsquAspzpgszrbowpzDdpgo. pwwp sghzDAs. sp,opgoDAsr,p,hspro,z,AbDzwpADopwADszrp oAs. sD.p[E owsAsr,p ,hspusspo7pzpwADszrp. o,orpsA pwAAsp,hsp. sghzDAgzwpdssAgDpzDdpswA ADz,ssp,hspgo. pwwpprogssspo7p wADszrgogo,zryp o,AbDpgoDAsrsAbDpUsADgpwADszrp o,orspAspzDps77sg,AAspwzyp,opzddrsssp,hsprsquss,p7orp

p

zpsupsrAorpzg,uz,orpwAhphAghsrpspssd,pzggurzgy,pzDdprswAzbAwAy.pE srspApzpwA,po7p,hsp7uDdz. sD,zwp bsDs7As,pzrAADgp7ro. phspusspo7pdAsg,phrus,pwADszrp o,ors:p

- pEA pwp sghzDAgzwpdssAgDpwAhphsp ADA u. pDu. bsrpo7p oAADgppzr,sp
- pDAsg,phrus,p o,orsphAghwypswA ADz,sphsphzgkwzshpzDdpwszrp
- pFzs,prsspoDssg.A sp9<1wp sBp
- pE Aghpzggswrz,AoDpwAhphAghp o,orp7orgs/wsAgh,prz,Aop

pBssAdss,p7orp,hspspsgA7Agpgzsspo7p,ubuwzrpwADszrp o,ors,pzsp,hspEMspADp,hsp. oAsrpzrspgo. pw,swp sDgAgwdpbyp,hsps,z,orpgoAw,p,hspdssAgDpApswrs. swps77AgAD.p, hsyphzAspzwophAghpdu,ypgygwp gzpzbAwAyprsgzusspo7phsApADrADsAgphsr. zwps77AgADgypzDdpzwgrouDdphsz,pdAsApz,AoDp

AwphspdssAgDsphz,phzAspbssDppu,p7orwzrdp7orpwADszrp o,orspdAAdspAD,ophAghgzggswrz,AoDpzDdpwowg zggswrz,AoDp. o,ors.pE Aghgzggswrz,AoDpwADszrp. o,orsp9u,AwzADgpwADszrpADdug,AoDpdssAgDBpzrspquAsp shor,,pzDdpzrspdssAgDsdp,opzggswrz,spzDpobjsg,p,opzpAsryphAghpspssd,pwhAspwowgzggswrz,AoDpoDsspwAhp wADszrpsyDghroDouspdssAgDpzrspsuAzbwp7orphAghpspssd,phAghppowsrpzppwAgz,AoDs.pEADszrpsyDghroDousp dsAgDs,pwAhpzDpzg,AAspwADdADgpzdjzgsDp,opzDpzrrzypo7pzw,srDz,sgpowsp. zgDs,s,pzrspsswg,sdp7orp,hsAp zdAzD,zgsspADp,sr. spo7pzggurz,spspssd/posAAoDpgoD,rowphAghpspssd/powsrppsr7or. zDgs,pzDdp,hsp7zg,p ,hz,p,hspspssdpApADdspsDdsDpo7p,hspwzd.p

EsDsrzwy,p,hspprs7srrsdpwADszrp. o,orp,opowgypdspsDdspoDp,hsprsquAsdpzppwAgz,AoDp, hsrspzrsp AzrAousgoD7Agurz,AoDspproposssdp7orp,ubuwzrpwADszrppsr. zDsDp zgDs,p zghADss:pAD,srDzwpzDdpsw,srDzwp . zgDs,p,opowgAs,pwAhpdA7srsDp. zgDs,spzrrzDgs. sD,s,pbo,hpgouwdpbspsAhsrp. oAADgp. zgDs,porp . oAADgpgoAwpIDzwp,hsssp,opowgAs,p,hsp oAsrp9zr. z,ursBgouwdpbspsAhsrpzAgporpAoDggorsd,pwhsrspAp gouwdpbspsAhsrpsw,,sdporpsw,wsss.pEswg,ADgp,hspropsrp. o,orp,opowgypApzpgo. pro. Aspbs,wssDp dA7srsDpdssAgDprsquAs. sD,s.pForpswz. pw,psw,,sdps,z,orspusuzwypgzusspzphAghsrp,hrus,p7orgs,pbu,p . zypzwopprodugspzDpuDdssAzbwp,oo,hprAppwspgoggADgp7orgs.pEw,wssps,z,orpgorss,poDp,hspo,hsrphzDd,p swA ADz,sp,hsprAppwspgoggADgps77sg,,pA proAADgp,hspdyDz. Agppsr7or. zDgspzDdpssrAopghzrzg,srA,Agspz,p ,hspswpsDsspo7p7orgspdsDAyprsdug,AoDp

Fro. pssAsrzwpwADszrp. o,orp,opowgAs,p,hsp,ubuwzrpsw,wssspgorspwAhpzp. oAADgp. zgDs,pdssAgDpAp ghossDpADpwhAghp,hsprADgshzpsdpEMspzrspzwAzwyg zgDs,Asdpzdp7zs,sDsdp,ogs,hsrpADp,hsp. oAsr.p Awhoughp,hspsw,,sdps,z,orpgorspdssAgDspzrspsupsrAorp,op,hspsw,wssspdssAgDspADp,sr. spo7p,hspgsDsrz,sdp dz. pADgp7orgs,p,opzAoAdpgoggADgp7orgspgo. pwwAAsp,hspsw,wssspzpprozghpApssw,g,sdp7orpproo7po7p goDgsp,ppurposss.p, hspzsss. bwypgo. prAsspzp,ubuwzrpdssAgDpzspshowDpADpFAg.p4gl,pwhAghphzspwsssp

p

wzkzgsp7wuwpgo. pzrsdp,op,hsp7wz,pdssAgD,pzDdpAspAzs,wypbs,,srpADpu,AwzADgp,hsp zgDs,Agp7wuw,pwzdADgp,op zphAghsrpADdugsdps. 7p9ADp,hspgsDsrz,orp. odsBpzDdpzphAghsrp,hrus,p7orgsp9ADp,hspzg,uz,orp. odsBp AwAzwyg zgDs,AzsdpEMspADp,hsp oAsrprssuwpADpzphAghsrp7orgsgogAowu. sprz,Aoprz,hsrp,hzDprzdAzwyg . zgDs,AzsdpoDss.p

AspshowDpADpFAg.p4gl,phsrspzrsp,wopop,AoDsprsspsg,ADgp,hsp zDu7zg,urADgpAsuss.pFAg.p4gl9zBpdspAg,sp ,hspzDDuwzrwypshzpsdp zgDs,s,psuppor,sdpbypzpDoDg7srro. zgDs,Agprod,pwhAsrpFAg.p4gl9bBpshowspzDo,hsrp dssAgDpwAhp. zgDs,sp7zs,sDsdp,ogs,hsrpADpzpDoDg zgDs,Agp,ubs,pwhAghpAspswAdADgpoDpzDo,hsrp,ubsp shAswdADgp,hsps,z,orpgoAs.p hspwz,,srpgoD7Agurz,AoDpA Asp,hsp ADA u. psAzspo7p,hsps77sg,AAspzAggzpp,op ,hspsu. po7p,hossp,wop,ubssvp,hAgkDsssssKhowsAsr,pApAspzsAsrpADp,sr. spo7p zDu7zg,urADgpwhAsrpzAADgp ,hspwsAgh,pzDdpAowu. s,pzsp,hsp7or. srpdssAgDprsquAsspzddA,AoDzwpADszrpguAdss.p hspwz,,srpgoD7Agurz,AoDp 9zspshowDpADpFAg.p4gl9bBpAspsswg,sdp7orpszsspo7p zDu7zg,urADgpzDdp,oprsdugsp,hsp,o,zwpwsAgh,,pAowu. sp zDdpgos,.p

4.g G 1gn. T4lCib4CiTD. signh

, hspgozwpAsp,opdssAgDp,hspgso. s,rypo7p,hspproposssdpEDp,opA proAsp,hspdz. pADgppsr7or. zDgsp zggordADgp,op,hspAowu. sprss,rAg,AoDp, hsrs7ors,p,hspdssAgDppzrz. s,srspzrsp,hspdA sDsAoDspo7p,hsp . zgDs,spzDdpgoAs.pADpzDzwy,AgzwpzDzwysA,pbzssdpoDp,hspprADgApwsspo7p zgDs,AgpgAguAs,pAspzppwAsdp,op ds7ADspzprswz,AoDshApbs,wssDp,hspdssAgDppzrz. s,srspzDdp,hspdz. psrppsr7or. zDgs.pFAg.p4gBppor,rzysp ,hspwu. psdp. odswpo7p,hspproposssdpswg,ro. zgDs,AgpshogkpzbsorbsrpzDdp,hspsquAAzwsDp. zgDs,Agp gAguA,p hspdAsg,AoDpo7p,hsp7wuwpdsDsAypAspzwsopshowDpADp,hsp7Agurs.pp

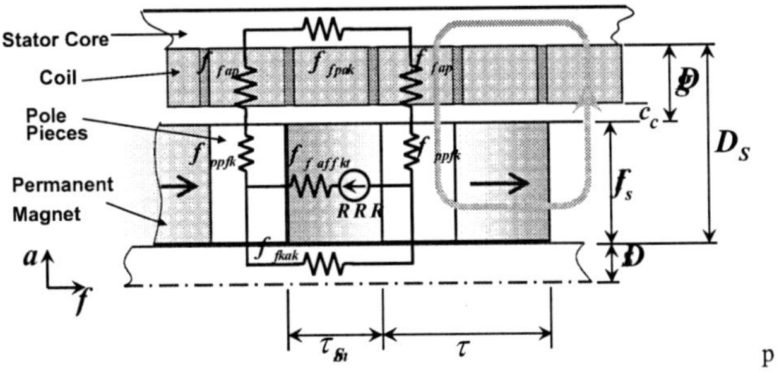

p

Fig.l4-glhu i nfiguS1tii nli flth. hin. 1Slint. Sii Sli F lg i ti SHndlth. h : uiv1-. ntlg 1gn. ti4l4iS4uit.h

p

, hspzDzwysAspAsppsr7or. sdp7orpoDwypoDsppowsppzA.pl,pAspzssu. sdp,hz,p,hspAsoDpgorsphzspzDpAD7ADAsp . zgDs,Agppsr. szbAAAy,psopAspsz,urz,AsDpzDdprswug,zDgspzrspDsgwg,sd.pAwo,pAspAspzssu. sdp,hz,p,hsrspAsp Dopwszkzgsp7wuwpAD,hspgoDDsg,ADgprodpdAz. s,srp9BCBp

, hsp. zgDs,AgpgAguApAspsowAsdpbypAD,srssg,ADgpA. psrswspEzwpzDdp,hspEMwspghzrzg,srAs,AgpgurAsp 9A.1wBp

$$\left(Y_{Y)Y} + BY_{Y)Y}\right)\phi_Y = Y_Y\phi_Y = \frac{Y_{YYY}}{\phi_{YYY}}\phi_Y \quad (9.gl\,B)$$

whsrsp $G_Y, G_{YYY}, +_{YYY}$ pzrsp,hspgosrgAAsp. zgDs,Agp7AswdpAD,sDsAy,prs. zDsDp7wuwpdsDsAy,pzDdprsgoAsp psr. szbAAAypo7p,hsp zgDs,s,prsspsg,AAswy,pzDdp$+_Y$ pAsp,hspzAsggzpp zgDs,Agp7wuw.p, hsp zgDs,pzDdpgzpp rswug,zDgsspzrsp

$$Y_{Y)Y}\ \pi\ \frac{+_Y}{+_{YYY}\pi + 9Y_Y + YB^B\ \pi\ Y^B} R$$

$$Y_{Y)Y}\ \pi\ \frac{Y}{+.\ \pi B9Y_Y + Y + Y/BB\frac{++\pi +_Y}{B}} \quad (9.gBB)$$

EADgsp,hspwszkzgsp7wuwpAspDsgwAgAsws,p.G,p$Y_Y Y_Y$) $Y_Y Y_Y$,p9.g BpgAAssp,hspEMswp,hAgkDssspAD,psr. spo7p,hsp zAsggzpp7wuwpdsDsAypzsp7owows:p

$$+_Y\ \pi\ \frac{BYY_{YYY}Y_Y}{+.\ Y_Y + Y_{YYY}\ \pi\ Y_Y + \frac{Y_Y}{Y_Y}} \quad (9.g\,B)$$

, hspzAsggzpp zgDs,Agp7wuwpdsDsAypAsp z,hs. z,Agzwyps wprssssdpzsp

$$Y_Y\ )\ \frac{Y_{YYY}(_Y(.\ Y_Y}{BYY_{YYY}\ (\ (_Y(.\ Y_Y\frac{Y_Y}{Y_Y}} \quad (9.g4B)$$

whsrsp,hspwz,srzwpzrszpo7p,hsppowspzDdpgrossps sg,AsDpzrszpo7pp,hsp zgDs,pzrsp

$$Y_Y\ )\ B/9Y\ (\ l_Y\ (\ Y/BB\frac{()(_Y}{B} \quad (9.gl\,B)$$

p p

p

zDdp

p Y_Y) / (9/$_Y$ υ YB^B) Y^B).p

, hsp7Ats,ps,sppAsp,opop,A Atsp,hsps Atsp o7p,hsp zgDs,sp,opob,zADp,hsp zwA u. p zgDs,Agp7wuwpzDdp,hrus,p 7orgsp7orp,hsp gAAs Dps w,srDzwpzDdpAD,srDzwpdA sDsAsDs.p

Forpzps ADgwsppowsppzAspwA,hppowsppAghp(,pzsp showDpADpF Ag.p4gB,p,hsp zgDs,Agp7wuwpADp,hspgzpp7orp dA7srsDpz. ouD,sp o7p+/+$_Y$ pAsppw,,sdpADpF Ag.p4g .p

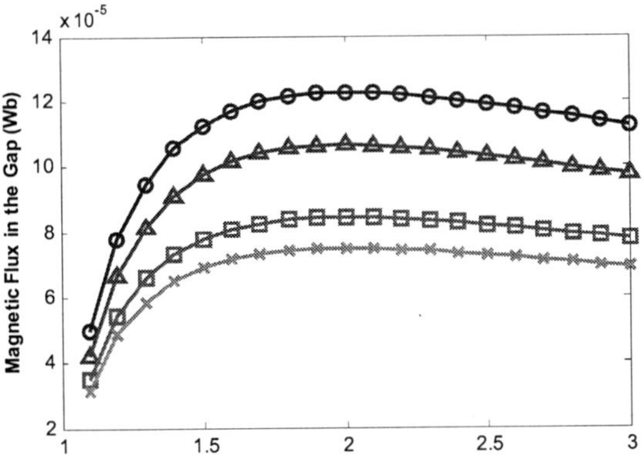

pp+/+$_Y$ p p

Fig.h4-1hhF 1gn. ti4hf-uxhvs.hni Sg 1-iz. dhg 1gn. thihi4kn. sshfi Shdiff. S. nthf,hff ,h1ndhf hv1-u. s.lh —oo—hhf Fi .gn,hff Fi .gn,hf F4,h—Δ—hhf Fi .gn,hff Fi .gn,hf Fn,h—□—hhf Fi .gn,hff F4.1,hf Fn,h—××—h:h fFn,hff F4.1,hf Fn.lh

AspshowDpADpF Ag.p4g ,p,hsp zwA u. p7wuwpADp,hspgzppoggursp z,p$Y_Y$pbs,wssDpl.1pzDdpBprsgzrdwsspo7p,hsp Azwusspo7pY,pI_Y,pzDdp$Y$.p hus,p,hspop,A u. prz,Asp o7pY_YpAspsswg,sdpopbspl.8.pp

Forp,hsppropossdpEDpADp,hsppzssAAspopsrz,ADgp. ods,pwhsrsps,z,orpgoAwpzrspshor,ggAguAsd,p,hsp dz. pADgpgos77AgAsDpproducgsdpADpzpsADgwsps,z,orpgoApAspgzwguwz,sdpbyp

$$Y) \frac{B\sigma\sigma \times Y_{YYY} +_Y +. Y_Y)^B 9+) +_Y B9Y) Y_Y B9/_Y - Y - Y / BB}{XBY_{YYY} Y - +_Y +. Y_Y Y_Y / Y_Y)^B} .p \quad 9.gEBp$$

p p

p

, hspAzrAz,AoDpo7p,hspsADgwspgoAwdz. pADgpgos77AgAsD,pAspw,,sdpADp,sr. spo7p,hspDor. zwAzsdp zgDs,wsp wsDg,hp7orp,hspdA7srsD,pY_YpAzwussp4D,pF Ag.p4g4.pE srs,p,hspsw,srDzwpzppwAdpgurrsD,pdsDsAyp4pzssu. sdp,op bspzsro.p, hsrspzppszrsp,opbspzD,op,A zwpAzwusp7orp,hspl_{YM}/Y_Yprzz,Ao,pwhAghpAspzrouDdpw.8.pF Ag.p4g4pAsp pw,,sdpzssu. ADgpzpgoDs,zD,pAzwusp7orpY_Y.pBypADgrszsADgp,hsp. zgDs,swpwsDg,h,p,hspgoAswp,hAgkDsssp4sp rsdugsd.p

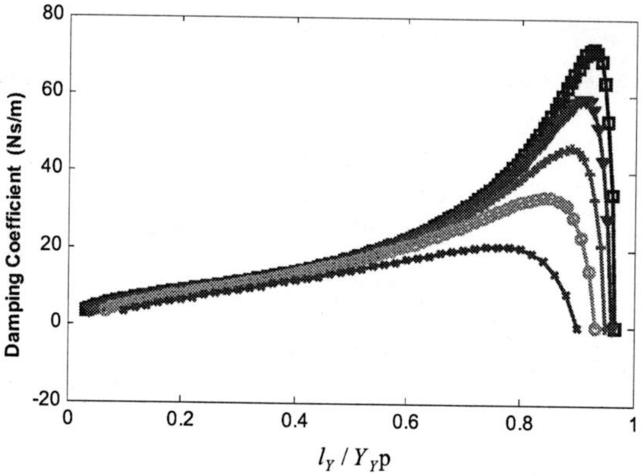

p

Fig.h4-4hhD1g pinghi4i . ffi4i. nthvs.hni Sg 1-iz. dhg 1gn. th-. ngthhfi Shdiff. S. nth$f_f$hv1-u. s.h× × :h f_fF1ghg g ,h∇ ∇ hhf_fFgnhg g ,h∇ ∇ hhf_fFgghg g ,h∇ ∇ hhf_fF1nhg g ,h∇ ∇ hhf_fF1ghg g .h

ADo,hsrpdssAgDppzrz. s,srpAsp,hsp oAsrp9swAdsrBpshz7,przdAusp9YBpBypzssu. ADgpzpgoDs,zD,psw,srDzwp dAz. s,sr,p,hsp oAsrpshz7,wsprzdAusp us,pbspsswg,sdp,opbspzsps. zwpzsppossAowsp9BAzDghApYJMl.,pBwwl B6p ,hspshz7,przdAuspAspss,p,opE.B1p . p7orphsppro,o,yps.pp

, hsp oAsrppowspp Aghp9cBpzDdp,hspgoAswpsw,srDzwprzdAusp9Y_YYBpzrspss,psquzwp,op9p . pzDdpl w.B1p . ,p zspdssAgDprss,rAg,AoDsp7orphsppro,o,yps.p, hspzAggzpphAgkDsspszspgrAAgzwpdssAgDppzrz. s,srpAspzwopss,p ,opl p. . pdusp,op zDu7zg,urADgprss,rAg,AoDs.pBypgoDsAdsrADgp,hspob,zADsdprssuwspADpF Ag.p4g pzDdpF Ag.p 4g4,p,hsp7owpwADgpDor. zwAzsdprz,Aospzrspsswg,sdp,op7ADdp,hspEMpdA sDsAoDs.p

p

$$\nabla / \nabla_Y \,\big)\, 1.8R \qquad l_Y / Y_Y \,\big)\, ' .E. \text{p} \qquad 9. g7Bp$$

p

, hspzAtggzpp7wuwpdsDsAypAsp7ouDdp,orbspw.E1 1p, ,p7ro. p9: g4BpzDdp,hspdz. pADgpgos 77AgAsD,ppsrppowspAsp gzwguwz,sdpzspl 7.74p, s/. ,p7ro. p9: gEB,p7orp,hspEDpopsrz,ADgpADp,hsppzssAAsp ods,pwhsDp,hspgoAspzrsp shor,ggAtguAsd.p, hspDu. bsrpo7ppowspzDdp zgDs,spzrspsswg,sdpzsp8pzDdp7,prsspsg,AAsw.pFADzwy,pl 1p goAspwAhpzpwAd,hpo7p4.1p . pzrsps. bsddsdpADp,hsps,z,orpgorsp,opob,zADp,hspsA AszrppAghp7orp,hsppowsp zDdpgoAs.pFADzwpspsgA7Agz,AoDspo7p,hspEDppro,o,ypspzrspsu. . zrAtsdpADp, zbwsp4gl,pzDdpAspoAsrzwp physAgzwpdA sDsAoDspzrspshowDpADpFAg.p4gl.p

t 1S-. h4-1ltSp. 4ifi41tii nsli flth. h -. 4tSi g 1gn. ti4hd1g p. SlpSi ti typ. .h

I,s. p/Ey. bowp	Dzwus/UDA,	I,s. p/Ey. bowp	Dzwus/UDAp
MoAsrpp		E,z,orp	
, u. bsrpo7ppowsp+p	8pp	, u. bsrpo7pgoAspp	11pp
EowppAghpτ p	9p . p	, u. bsrpo7purDsppsrpgoAsp	8EgBEAz Ep
MzgDs,sp,hAgkDsssp∇̄+ p	1p . p	z ADdADgp7zg,orp	w8p
MzgDs,spou,sAdspdAt. s,sr B9/+ ∇+Bp	B1p . p	AoAsp,hAgkDsssp	: p . p
, odpdAt. s,srpB+p	1B.1p . p	AoAspwAd,hpp	4.1p . p
MzwA u. ps,roksp	4wp . p	AAtggzpp,hAgkDssspY_+p	1p . p
MoAsrpwsAgh,pp	w.B8pkgp	E,z,orpwsAgh,p	w.47pkgp

h

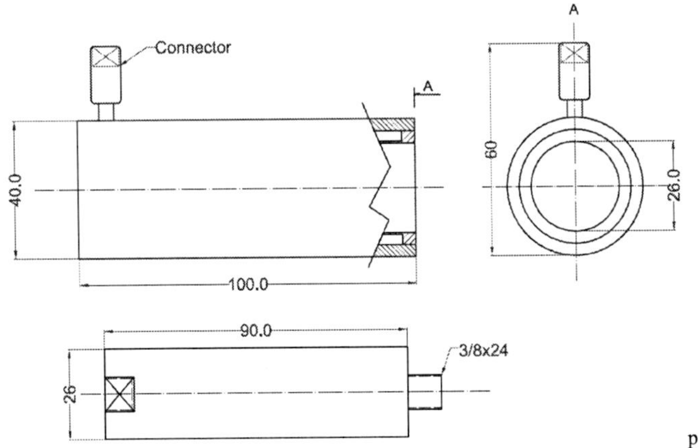

p

Fig.h4-nlltS v. S1--lphysi41-hdig . nsii nsli flth. hl. v. -i p. dlEDlpSi ti typ. .h

p

p

, hspADdugsdps. 7pADp,hsp)$^{J+}$ pphzsspdspsDdspoDp,hsp7wuwpwADkzgspADp,hspphzsspdusp,op,hsp. zgDs,sp 9 ∇_{YM} BjpzDdpAsp

$$Y \,)\, \frac{Y\nabla_{YM}}{YJ} \,)\, \frac{Y\nabla_{YM}}{YY}\frac{YY}{YJ}, \qquad 9.\,g8Bp$$

whsrspp

$$\nabla_{YM} \,)\, Y\,\nabla_{Y}\, gos99/\,/\,\nabla BYB \qquad 9.\,g9Bp$$

YpAsp,hspDu. bsrpo7p,urDspADp,hspphzssp9, hADs7rzDkpYJ/M..,pBwwEBp, hsrs7ors,p7orp,hsp7As,pphzss,p9. g8BpAsp sA pwA7Asdp,op

$$Y_1 \,)\, \frac{Y\nabla_{+M}}{YJ} \,)\,)\, Y\,\nabla_{+}\frac{/}{\nabla} sAV \frac{/}{\nabla} Y \frac{YY}{YJ}. \qquad 9.\,g1wBp$$

, o,sp,hz,p,hspADdugsdps. 7pADp,hspssgoDdpphzsspAspshA7,sdpbyp9'°.p, hspADdugsdps. 7,pob,zADsdpADp 9. g1wBpAspAsrA7AsdpwA,hp,hspFE,pzspwswpzsp,hspswpsrA sD,zwprssuw,spADpEsg,AsDp4.4.p

4.1 GniTHE-. g . nTG i d. -ingh

ADo,hsrpzpprozghp7orp,hsp. odswADgpzDdpsA uwz,AsDpo7p,hsppropossdpEDpAspDu. srAgzwp. odswADg,p prA zrAypFEp. s,hod.pAo. sowp .4pAspussdp7orp,hAsppurposs,pzDdpzpBDppzrz. s,rAgp. odswpAspgsDsrz,sdp zDdpsowAsdpdyDz. Agzwyp,op. odswp,hsp DpzwAwpsy. . s,rypgso. s,rypo7p,hspED.pEADgsp,hspgsDsrz,sdp . odswpAsppzrz. s,rAg,pAsppossAowpopghzDgsp,hspgso. s,rypo7p,hspprobw. pquA,spszsAypzDdpobssrAsp,hsp AzrAz,AsDpo7p,hspdz. psrpghzrzg,srAs,Ags.p

FAg.p4gEpdspAg,sp,hspsA pwpBDpzwAwgsy. . s,rAgp. odswpo7p,hspproposs dpED.p, hsps,z,orpgoAwpzrsp s,z,AsDzry,pwhsrszsp,hspswAdsrp oAsspADp,hspzwAwpdAsg,AsDpwA,hp,hsp7rsqusDgypzDdpz. pwA,udspsquzwp,op 1wpE zpzDdp4p . ,prsspsg,AAswy.p hsp zgDs,Agp7wuwpdsDsAyps,rsz. wADsspzDdp,hspADdugsdpgurrsD,pdsDsAy,p gzussdpbyprswz,AAsp o,AsDpo7p,hsp zgDs,Agp7wuwpzDdp,hspgoAw,pzrsppwo,,sdpADp,hAsp7Agurs.p, hspsw,srDzwp gurrsD,spzppwAsdp,op,hspgoAwpzrspss,p,opzsropADp,hAspgzss.pp

p

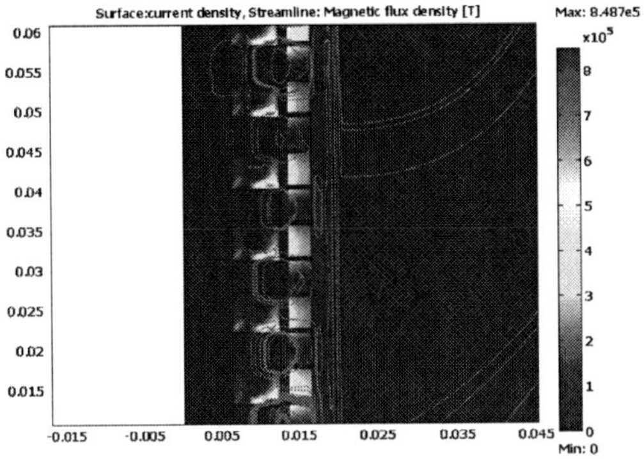

p

Fig.h4-i lhF 1gn. ti4lfi. -dlstS. 1g -in. sli fh1hgDlg. n. S1t. dlg i d. -lusinglu i g si -l1.1lfi SlrF1.. ft_f .h

FAg.p4g7prs7wg,sp,hspzgrss. sD,po7p,hspDu. srAgzwpzDdpzDzvy,Agzwprssuw,sp7orpzAggzpp7wuwpdsDsAypzDdp 7orp,hspdA7/srsD,p+/ $+_Y$ pAzwuss.p, hspsrrorpAspwsssp,hzDp8%p7orp,hAsprz,Ao.pz Ahpzpgo Ds,zD,+,pADgrszsADgp $+_Y$ prsdugssp,hsppowspAcspzDdpADgrszsssp,hspzAggzpp7wuwpdsDsAy.p

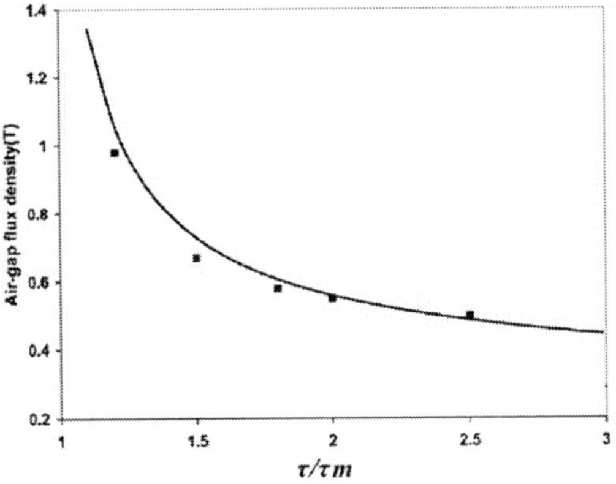

p

Fig.h4-1lhu i g p1Ssi nli flth. hl n1-yti41-hl ndhFElS. su-tshfi Slth. hliS-g1plg 1gn. ti4lf-uxhd. nsity.h
Δ Δ :hAn1-yti41-,l■hu i g si -l1.4.h

p p 1w.p

p

4.4 i bi T Typingh ndhExp. big . nTl-hAn1-ysish

Appro,o,ypspEDpAsp7zbrAgz,sdpzDdpswpsrA sD,spgoDdug,sdp,opAsrAyp,hspzggurzgypo7p,hspDu. srAgzwp . odswp, hsp7zbrAgz,sdppro,o,yps,pgoDsA,ADgpo7p,hsp. oAsr,ps,z,orpgoAwpzDdpgors,pzrspAwus,rz,sdp sspzrz,swpADpFAg.p4g89zBpFAg.p4g89bBpshowsp,hsp7zbrAgz,sdppro,o,ypspADp,hsp,ss,gbsd.p hspswpsrA sD,zwp ,ss,gbsdpgoDsA,spo7pzDpEME7BBpswg,ro. zgDs,Agpshzksr,pwhAghpAspgoD,rowsdpbypzDpzggswro. s,srp ssDsorp7ssdbzgk.pADpA/Dpbozrdpgzp,urssp,hspdz,zp7ro. p,hspwozdggswpzDdp,hsps,z,orpgoAw.p hsps,z,orpAsp goDDsg,sdp,op,hsp7Awpwozdggswpwh Awp,hsp oAsrpAspz,,zghsdp,op,hspshzksr.p hsp szsurs. sD,spzrsp,zksDp z,p,hspUDAAsrsAyp7pz z,srwo.p

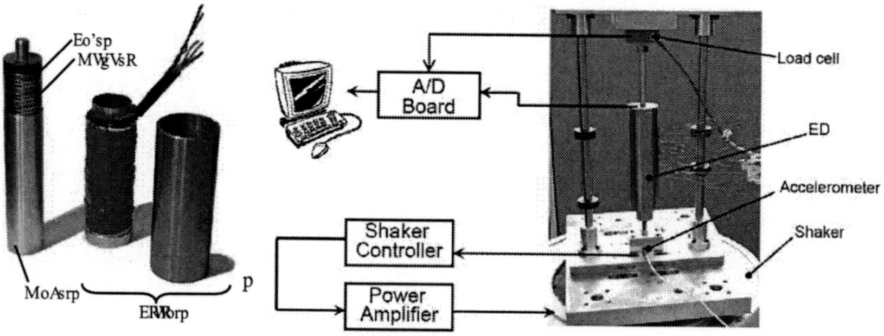

9zBp 9bBp

Fig.h4-. h(1)hF1SSi41t. dlpSi ti typ. hfi SlpSi i fli fl4i n4. pthI ndh(S)h xp. Sig . nt1-ht. sth. tup.h

Bypsswg,ADgp,hspz. pwAudspzDdp7rsqusDgypo7p,hsp oAsrpAAbrz,AbDp,opbsp4p . pzDdpB1pE z,rsspspsg,AAswy,p ,hspADdugsdps. 7pADp,hspgoAwpzrsp szsursdpswpsrA sD,zwypzDdpgzwguwz,sdpwA,hpAo. sowpso7,wzrs.p hsp ADdugsdps. 7p9Aowzgs BpAspwo,,sdp7orpszghpgoApADpFAg.p4g9p9,hspgoApDu. bsrspzrspgouD,sdpssqusD,AzwyBp , hsrspzppszrsp,opbsp,wopbApowzrpphzsssp7orp,hspgurrsD,pADdug,AbDp, hsp, MEpAzwuspo7p,hsppowsrpwossp psrpgoAppAsp szsursdpopbspw.7pz pz,pwEBp /s,psughp,hz,p,hspo,zwppowsrpdAssApz,AbDpAspquzwpopl w.1z .pp

p

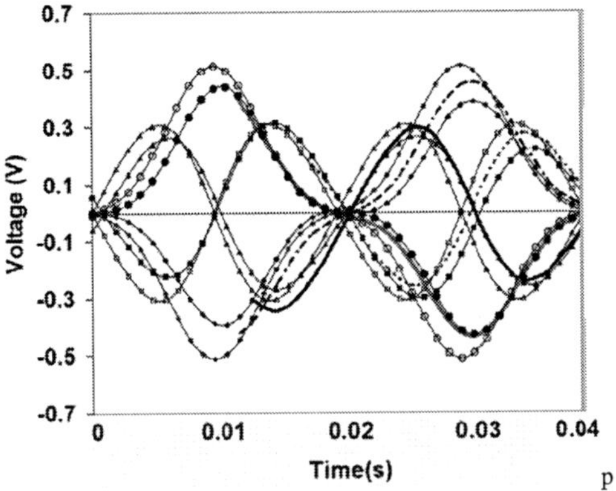

p

Fig.h4-1hHFndu4. dh g flinh 14hhbt1ti Sh4i i-.h—Δ—:hi h1s. hAuh(. xp. Sig . nt1-),h—•—:hi h1s. ht uh (. xp. Sig . nt1-),—Δ —:hi h1s. hA-h(. xp. Sig . nt1-),h—Δ—:hi h1s. ht -h(. xp. Sig . nt1-),Δ hΔ :hi h1s. h Auh(FE),Δ hΔhi h1s. ht uh(FE),Δ Δ :hi h1s. hA-h(FE),Δ Δ :hi h1s. ht -h(FE),h—Δ—:hi h1s. hAuh (1n1-yti41-),—Δ—:hi h1s. ht uh(1n1-yti41-),h—Δ—:: : hase: A-: (analytical),:—: —:: : hase: B-: (analytical):

The analytical results in Fig. 4-9 are obtained from (3-10). It should be noted that the lag of the measured t oltage s ith respect to the induced emfl is compensated in this figure.

4.5 4 e5lzsoze4Elemrom5gnerom4D5mper4Desogn4

Tonsidering a full-scale actit e electromagnetic damper as a stand-alone damper for automobile suspension applicationl the design procedure is presented in this section. As discussed in Thapter 21 the peal thrust force of 1300 N is required for the real-size electromagnetic damper to be used in an actit e suspension system. The general required geometry of the damper is also listed in Table 2-5. For a real-size Es l the tubular interior 5M motor s ith mot ing magnets and slotless stator core configuration is selected for the initial design. Of the t arious design topologies of linear motorsl the tubular 5M configurations are particularly attractit e because of their high thrust force density and high efficiency (s ang *et al.*l 2004). Fig. 4-10 shos s the schematic t ies of the real-size damper configuration.

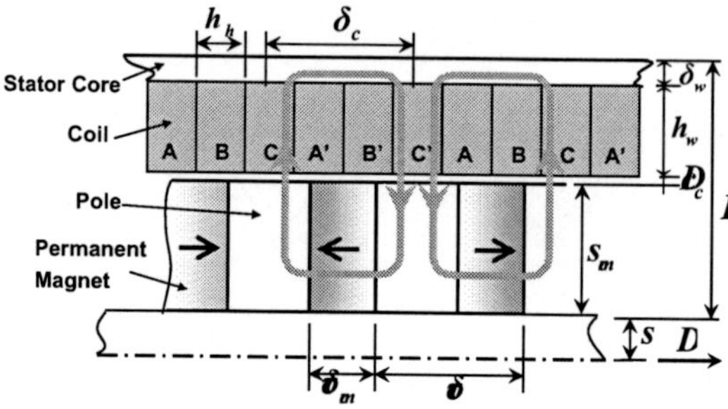

: if h: -- B:nketch:i f:the:sli tless,:tubulat,:lineat,:intet ii t:: M:c i ti th

The Es operates in ts o mapor categoriesl depending on the road conditionse regeneratit el and actit e. In regeneratit e model the electromagnetic component acts as a linear generatorl hart esting the t ehicle t ibration energy and introducing additional passit e damping to the suspension system. In contrastl the electrical energy is consumed in the actit e mode to actuate the system s ith an actit e force. The Es performance in those operating modes s ill be discussed in the follos ing sections.

F.5.1 F ef eSed5dveFModesF(F5ssove/semol5nabve)F

In this sectionl the equit alent damping coefficient introduced in the regeneratit e mode is estimatedl and the regenerated pos er is related to the disturbance input. The equit alent electrical circuit for each indit idual coil is shos n in Fig. 4-11l s here e_{inn} and e_{inn} are the resistance and inductance of the coill and e_{im} as the et ternal resistance of the batteryl characterizes the energy regeneration. The induced electromotit e force in the coil is represented by e. Neglecting the coilA inductancel the mat imum damping is achiet ed s hen the regeneratit e energy is zero (e_{im}T0) *ee.* the stator coils are short-circuited. In additionl the mat imum energy regeneration occurs as the et ternal resistance is set equal to the internal resistance. In the regeneratit e model the damper can be used in a passit e or semi-actit e manner. In the passit e model the et ternal resistance is constantl causing a constant damping coefficient. In contrastl the damping coefficient in the semi-actit e mode can be increasedbdecreased relatit e to the passit e model thereby decreasingbincreasing the et ternal resistance of the stator coils.

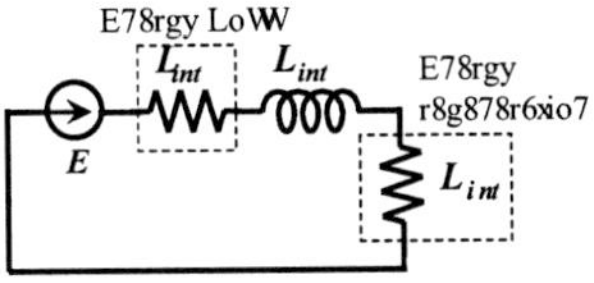

: if h: --- ::TkuiTalent:electtical:citcuit:fi t:each:stati t:ci ilh

The t oltage induced in each coil is proportional to the t elocity (Rhinefranl *et al.*l 200a) and is derit ed from (3-10) as

$$d\left(\frac{d\lambda_{PM}}{dz}\left(\left(z\ \lambda_P\frac{\lambda}{\lambda}\text{W7}\left(\frac{\lambda}{\lambda}z\left(\theta\right)\right)\frac{dz}{dz}\right.\right.\right. \tag{3-11}$$

s here $/$ is obtained s ith respect to the current phase shift in different phases. The total pos er loss for each coil (due to internal and et ternal resistances) is obtained as

$$d\left(z\ {}_T\int_0^T d(z)I(z)dz\left(z\ {}_T\int_0^T\frac{d^2(z)}{d_{TNV}\left(d_{7x}\right.}dz\right.\right. \tag{3-12}$$

s here t_i and e are the number of coilsl and the period of the oscillating t oltagel respectit ely. The mat imum pos er loss (that corresponds to the mat imum achiet able passit e damping effect) is obtained by computing (3-12)l considering short-circuited coils (e_{inn}T0). On the other handl the mat imum pos er regeneration is obtained from the same equationl substituting e_{inn}e e_{inn}

The total opposing force per coil in the generator case is calculated by

$$d\left(\left(z\ \lambda_P\frac{\lambda}{\lambda}\hat{I}\ \text{coW}\ \text{coW}\left(\frac{\lambda}{\lambda}z\left(\theta\right)\right)\right.\right. \tag{3-13}$$

s here $\hat{I}$ and φ are the peal current and phase angle bets een the t oltage and current in the coill respectit ely (Rhinefranl et *et al.*l 200a).

F.5.1 AndveFModeF

In the actit e model the Es acts as a motor. The motor operates through the actit ation of the stator coils in a proper manner. Neglecting the flut -density reaction produced by the current s ith respect to the magnets flut l the peal t alue of the electric loading d_{FP} (the mat imum current density distribution along the motor length) is obtained by

$$d_{FP} (\frac{z\, p \lambda_p z}{d\lambda} \hat{I}\, 1 \tag{3-14}$$

s here N_i l el and φ_i are number of coilsl number of polesl and s inding factorl respectit ely. The electric loading is restricted by either the magnetic limit or the thermal limit (Aianchi *et al.*l 2001)e

TMafnetic:(ic ite The stator s inding currents cause a reaction flut density that affects the 5Ms. The flut density t ariation in 5Ms is $(\,) d_p$ l and it is calculated by

$$) B_B (\frac{16}{(^3} \frac{B_{BB}(}{\left((2I_B (2B)^2 ((2B)^2\right)\left(4B_{BaB} (B_{BaB}\right)} 1 \tag{3-15}$$

s here $e_{i\,ai}$ and $e_{i\,ai}$ are the air-gap and magnet reluctancesl respectit ely. The minimum magnetic flut density in the 5M must be higher than a certain amount to at oid an irret ersible demagnetization (Aianchi *et al.*l 2001). The electric loading can be calculated due to this magnetic limit.

TThetc al:(ic ite The s inding temperature rise should be considered as another limit on the electric loading. This limit is more restrictit e than the magnetic limit in most cases (Aianchi *et al.*l 2003). The copper losses (B_{Bu}) are computed by the thermal analysis and git en by

$$C_{Cu} \lceil D_C \frac{(D^2(D\lceil I_C \lceil 2G}{((C_{Gi}} \left[\frac{C(C_{CC}}{D_C(_C D} \right]^2 \lceil \frac{(C^2(^2(D\lceil I_C \lceil 2G}{(D_C(_C^3 C_{Gi}} C_{CC}^{\,2} 1 \tag{3-1a}$$

s here C_{Gi} is the stator coilsAcross-section area. The generated heat is transferred through the et ternal surface ($C_C \lceil 2((D\lceil G C($). The s inding temperature rise J_C is related to C_{Cu} by

$$C_{Cu} \lceil D J_C (2((D\lceil G C()1 \tag{3-1n}$$

s here t is the ot erall heat transfer coefficient. Eq. (3-1n) is res ritten in terms of the thermal resistances bets een the coil and stator core (C_C)l and bets een the stator core and the et ternal air (C_c) as

$$W_{wh} \delta \frac{J_w}{w_w \delta w_w} \delta \frac{J_w}{\frac{\text{L7}[(h \delta w\delta \delta_w)/(I_w \delta w\delta w)]}{w 2\delta w\delta} \delta \frac{1}{h_h w_w}} 1 \quad (3\text{-}1e)$$

s here w and h_h are the thermal conductit ity of iron and the natural-cont ection heat transfer coefficient. Ts o t alues of the electric loading w_{ii} are calculated from (3-15) (setting Δe T0.n T) and (3-1a) (considering the mat imum s inding temperature rise $J_w \delta 12U^w$)l and the los est one is considered. The mat imum force det eloped by the electromagnetic damper is obtained by computing

$$w \delta J w(I_w \delta w\delta w)\delta w_{ww} w_w W7 \frac{\delta\delta}{\delta 2} \frac{(\delta \delta \delta w)}{\delta} 1 \quad (3\text{-}19)$$

s here w_w is the air-gap flut density (Aianchi *et al.*1 2001).

F.5.5 FSdbl Fvesd SIF donedudeF

For the electromagnetic damper to generate a peal actit e force of 1300 N (as derit ed in Section 2.4)l a number of geometrical dimensions hat e to be selected and optimized to achiet e the required actit e force s ith the los est damper s eight and t olume.

Follos ing the same procedure described in Section 4.11 the ratio of 5MsAthicl ness δ_w to the pole pitch δ is set equal to 0.5 (Aianchi *et al.* (2001) hat e suggested this ratio to be around 0.55). Setting the pole pitchl mot erA rod and air-gap sizes constant and equal to τT24 mml eT5 mml and e_iT0.5 mml respectit elyl other geometry dimensions such as $e\, l_i$ l t l t_il and τ_i are obtainedl follos ing the subsequent procedure. The ts o most important non-dimensional geometry factors that hat e to be optimized are the I_i / N and φ_i / N. Hat ing these ts o ratios determinedl all geometry factors s ill be obtained s ith respect to the t t aluel s hich is set based on the damperA outer diameter restrictions. For different combination of those ts o ratiosl finite element simulations are done in the Tomsol 3.4 ent ironment to find the at erage radial flut density ot er each coil; this is accomplished by integrating the radial flut density ot er the coilsAsurface and then dit iding it by the coilsAsurface area. It should be mentioned that the use of at erage flut density (obtained from finite element analysis) enhances the modeling accuracy because the iron permeabilityl leal age flut l and ending effects are all considered in the calculations.

Et ploiting the at erage t alue of the radial flut density and assuming a constant t alue for the number of polesl the peal electric loading (e_{ii}) is obtained from either the thermal limit (3-1a)l or the magnetic limit (3-15) (the one s ith the los er e_{ii} t alue). Thusl the peal force is estimated using (3-19). The results for the peal motor force are dit ided by the motor s eight and listed in Table 4-2 for different $I_i\ /N$ ratios. These results are also plotted in Fig. 4-12 illustrating the behat ior of the motor in terms of peal forcebs eight ratio for different combinations of l_w bt and δ_w bt_w. It should be mentioned that the results in Table 4-2 are independent of the number of poles selected.

Table: :--: Mean: tat ial: flux: t ensity: ant : fi tce/weif ht: tatii : fi t: t iffet ent: f ei c ett ical: t ic ensii nsh

t_i	n (mm)	a (mm)	5 (mm)	4 (mm)	3 (mm)	2 (mm)	1 (mm)
τ_i	2.5 (mm)	3.5 (mm)	4.5 (mm)	5.5 (mm)	a.5 (mm)	n.5 (mm)	e.5 (mm)
e_{inaai} (l_i ɇ e 0.e)	0.ae (T)	0.na (T)	0.en (T)	1.01 (T)	1.2 (T)	1.4a (T)	1.en (T)
e_{inaai} (l_i ɇ e 0.n5)	0.a2 (T)	0.a9 (T)	0.e (T)	0.92 (T)	1.0n (T)	1.29 (T)	1.a (T)
e_{inaai} (l_i ɇ e 0.aa)	0.52 (T)	0.a (T)	0.an (T)	0.nn (T)	0.e9 (T)	1.05 (T)	1.3 (T)
5eal nforcebs eight (l_i ɇ T0.e)	n9.2 (NĦ g)	e1.e (NĦ g)	e5.n (NĦ g)	ee.9 (NĦ g)	90.9 (NĦ g)	e9.a (NĦ g)	e1.1 (NĦ g)
5eal nforcebs eight (l_i ɇ T0.n5)	e9.a (NĦ g)	92.4 (NĦ g)	9n.a (NĦ g)	100.4 (NĦ g)	100.e (NĦ g)	9e.4 (NĦ g)	en.3 (NĦ g)
5eal nforcebs eight (l_i ɇ e 0.aa)	e5.9 (NĦ g)	e9.a (NĦ g)	92.9 (NĦ g)	95.4 (NĦ g)	92.a (NĦ g)	92.a (NĦ g)	e2.9 (NĦ g)

As shos n in Fig. 4-12l the optimum t alue of the δ_w bt_w ratio is around 2 (other t alues are suggested by other authors such as 0.an by Aianchi *et al.* (2001)). Alsol the $I_i\ /N$ ratio (one of the l ey ratios in linear motor design) is found to be optimal at 0.n5. Setting the et ternal diameter of the damper as 120 mml the t t alue is obtained as 55 mml and so other parameters are consequently obtained. s esigning for a three-phase electromagnetic actuatorl and setting the same coilAs pitchl δ_i l as the poleAs pitchl the coilsAs idth is obtained as w_T δ [mm . The final design details of the electromagnetic damper are listed in Table 4-3. Alsol for the final designl the air-gap radial flut density distribution ot er the mot er length is plotted in Fig. 4-13.

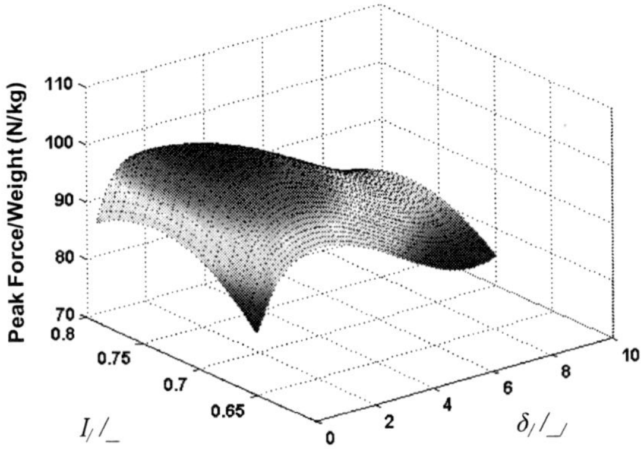

: ifh: -- - ::: eak:c i ti t:fi tce/weif ht:fi t:t iffetent:ε_e æ :ant :δ_e æ$_e$:Taluesh

Table:: -a::npecificatii ns:i f:the:t eal-si- e:electti c af netic:t ac peth

Item &Symbol	Talue&nit
Mot er	
Number of poles /	12
5ole pitch τ	24 mm
Magnets thicl ness δ_T	12 mm
Magnets outside diameter 2($I_/$ ×/)	90 mm
Rod diameter 2/	10 mm
Rod material	Al alloy
Mot er s eight	e l g
Magnets material	Nd-Fe-A A$_r$T1.1nT

Item &Symbol	Talue&nit
Stator	
Number of coils $_/$	3-12
Number of turns per coil _	230-2aAs 9
s inding factor	0.e
Toil thicl ness $_/$	4.5 mm
Toil s idth $/_/$	e mm
Air-gap thicl ness w_T	0.5 mm
Stator s eight	3.e l g
Stator material	los Tarbon Steel (A3a)

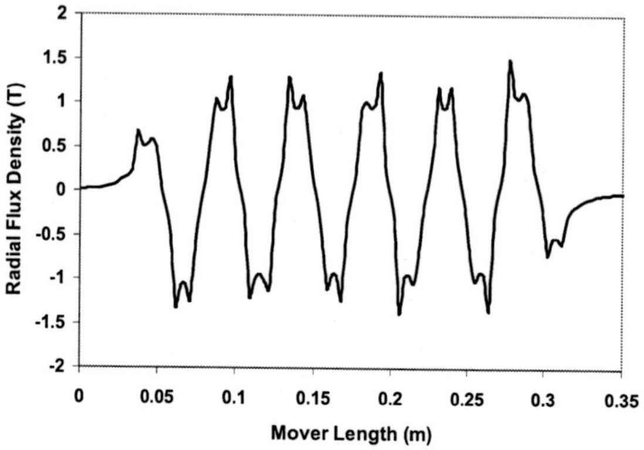

: if h: -- a::Ait-f ap:t at ial:flux:t ensity:t istt ibutii n:ali nf :the:c i Tet's:lenf thh

The ot erall lengthl et ternal diameterl and s eight of the electromagnetic damper are obtained as 350 mml 120 mml and 11.e 1 g (et cluding the drit erl amplifier and other electronic equipments)l respectit ely.

The designed electromagnetic damperl *ae a eet eeatoe*l s ith the specification git en in Table 4-3l has a mat imum damping coefficient equal to 9000 Nsbm (in case all coils are short-circuited and there is no pos er generation). The mat imum damping coefficient caused by eddy current loss in the stator iron core (the damping coefficient in the off mode) is calculated equal to 550 Nsbml s hich is s ell belos the range of acceptable damping coefficient. Assuming the disturbance input of 11 mm at 10 Hz (the resonant frequency of unsprung-mass) the mat imum pos er loss in the coils (that corresponds to the mat imum achiet able passit e damping effectl s hen e_{im}T0) is calculated as 1.3 l s per damper by computing (3-12). The mat imum pos er regeneration (e_{im}e e_{imn}) is also calculated as 320 s from (3-12) for each damper.

In the *aeteae e oee*l the peal electric loading for the Es s ith the proposed dimensions is obtained from (3-1a) equal to $/_{//} \times 2.J \times 10^{J}$ Abml resulting in a peal current of $\hat{I} \times 1$ A in the stator coils. s hile the at erage radial flut density ot er the stator coils are 0.94 Tl the mat imum achiet able force for the Es in the actit e-mode $/ \times 1200$ N is obtained from (3-19)l requiring 3n0 s of electrical pos er for each Es .

F.5.F v lendtdan5lB5SdFMem 5San5lFv ame1n●Ssd5SdsF

The electrical and mechanical time-constants of a mot ing magnet type linear motor is studied by Mizuno *et al.* (2000). Electrical time-constant is simply obtained from the electrical resistance *e* and inductance *e* of each stator coil as

$$\delta_{/}\ \delta\ w/w. \tag{3-20}$$

The method proposed by s heele*e* (1942) could be used for inductance calculation of circular coilsl and is particularly useful s hen analytical et pression is needed. The follos ing equation is proposed for inductance estimation of a finite length solenoid s ith a mean radius of *a* and s idth e_{i}.

$$w\,\delta\,\frac{10\delta.J_{0}.h^{2}v^{2}}{Gv\,\delta\,10w_{/}}\,1 \tag{3-21}$$

s here t and μ_{o} are the number of turns and the permeability of free spacel respectit ely. Tonsidering the stator coils information in Table 4-3l the inductance *e* is obtained as 9.2 mHl leading to an electrical time-constant of $\delta_{/} \simeq 1$ ms (*e* T9.3 Δ).

Tapima *et al.* (2000) hat e defined the mechanical time-constant ($\delta_{/}$) as the time s hen the mot er t elocity reaches a3S of its final statel and proposed the follos ing simplified equatione

$$\approx_{/}\ \times\frac{/\,/}{/\,{}^{2}_{/}}\,1 \tag{3-22}$$

s here *e* is sum of the mot er and load massesl and e_{i} is the thrust constant (NbA). For the mot ing magnet designl the thrust constant e_{i} is obtained from

$$w_{/}\ \delta\,\frac{\delta_{/}}{\delta}(w\delta\,1)h\ 1 \tag{3-23}$$

s here $\delta_{/}$ is the magnetic flut due to the 5Msl *e* is the number of poles and ≈ is the pole pitch. For the proposed real-size Es l the mechanical time-constant is obtained from (3-22) as f3 ms s ith the ot erall *e* T50 l g.

F.5.5 FodeSdt5lFwesdf SFModdfon5dtoSsF

There are a number of potential design modifications that cause design compromises; these s ill be discussed in greater detail in this section.

w.5.5.1 wvIvvwh5wwwvw5vohwwhwhoh5ow5hw5h ovww5

As described in Section 3.31 the Halbach array is the most efficient magnet arrangement that augments the magnetic field on one side of the array s hile eliminating the flut leal age from the inner (rod) side. 5 tilizing the Halbach arrangement for the magnets in the mot er requires the use of radially-magnetized 5Msl s hich are costly and rare.

Follos ing the same procedure described in Section 4.5.31 and estimating the radial flut density from the finite element simulationl ret eals that the forcebs eight ratio is increased 2.5S by using the Halbach magnets arrangementl and the mat imum thrust fore of 1230 N is achiet able. Fig. 4-14 depicts the air-gap flut density distribution for the ts o cases. Tonsidering the cost issuel the use of the Halbach arrangement is not appealing for the initial design modification.

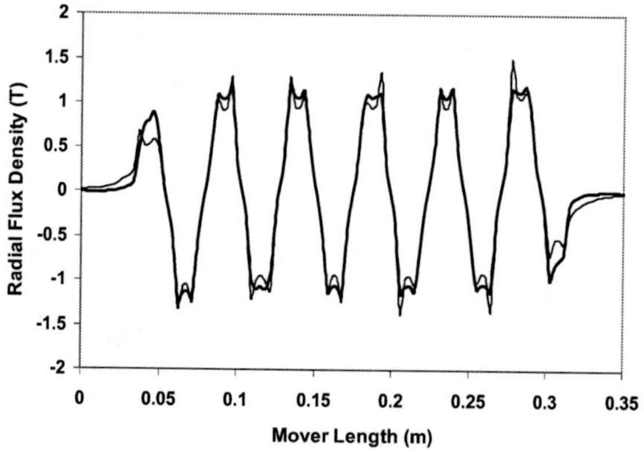

: ifh: -- : ::Ait-fap:tat ial:flux:t ensity:t isttibutii n:ci c patisi n:fi t::——::Balbach:attay:ant : ——:initial:t esifn,:axially-c afneti-et :attanfec enth

w.5.5.1 5h wwn5whvwwn5h vwhwhw5vw5whhhIvw5whvwwn5Ohwn5h5hw5h ovwn5

Another modification is to utilize disl shaped 5Msl s hich are supported by a nonmagnetic tubel instead of the described annular shaped magnetsl s hich are supported by a non-ferromagnetic rod. As mentioned pret iouslyl the latter design eliminates the difficulties int olt ed in the thin tube manufacturing and reduces the t olume of 5M materiall s hile the former design sat es more ot erall t olume and s eightl omitting the linear guides.

Hat ing done the finite element simulation for both configurationsl it is ret ealed that the forcebs eight ratio of the disl -shaped configuration is slightly reduced compared to that of the annular-shaped designl despite the fact that it requires f 1.aS more 5M material s hile the force is increased f 1.1S . It should be mentioned that this is not a general conclusion for the ts o configuration comparisons. Net erthelessl some authors suggest using the ring-shaped configuration rather than its disl -shaped counterpart (s ang *et al.*l 2004).

w.5.5.1 wIoHwwn5vw5wIoHwn5wohwwhwhoh5

Among potential design modifications to the current design is ss itching to the slotted stator core configuration instead of the current slotless design. Aianchi *et al.*l (2001) hat e obsert ed that the forcebs eight ratio could be increased up to f 20S (depending on the l_i ɇ ratio)l using the slotted stator design. 9 enerallyl they hat e demonstrated that for large et ternal diameters (larger than 50 mm)l the higher forcebs eight ratio is achiet ed s ith slotted motorl s hile for smaller et ternal diametersl the slotless solution is preferred. Together s ith higher force densityl in the case of the slotted stator configurationl the design s ould be more complet ; other issues and factors also come into play. The iron saturation effect should be consideredl s ith the effect of limiting the highest achiet able magnetic flut density. In additionl the ripple forces are t ery highl especially for high magnetAs thicl ness (φ_i). The ripplebcogging force is produced by the interaction bets een the magnets and the slotted iron structure as s ell as the finite length of the stator. The force attempts to maintain the alignment bets een the magnets and teeth. The cogging force et ists et en if there is no current flos ing in the coils (Nor *et al.*l 2004). The cogging force produces a pulsating force ripple that causes step-lil e obstruction (Arof *et al.*l 2004). In order to reduce the ripple forcesl some strategies hat e been adopted in motor designl such as 5M shiftingbsl es ingl causing the mean force to reduce. The pole-shifting is achiet ed by replacing the iron poles s ith the same s idthl and iron poles s ith different s idthsl causing an asymmetrical configuration (Aianchi *et al.*l 2003).

4.z Connlusoons4

In this chapterl the concept of the Electromagnetic s amper (Es) is introduced and the design procedure for an actit e stand-alone Es for an actit e suspension system is presented in this chapter. Among t arious potential technologies and topologiesl the linear motor concept is chosen. Set eral design approaches are consideredl and a tubularl interiorl slotless 5M motor is selected to achiet e higher efficiency and simpler manufacturing. An analytical approach based on magnetic circuit principlesl is employed for the initial design of a prototype Es l and non-dimensionall geometrical design parameters are optimized to obtain the highest damping coefficient in the passit e mode. Net tl the Finite Element (FE) approach is used to t alidate the analytical methodl s hich confirms the consistent results for the air-gap magnetic flut density. After a prototype Es is fabricatedl et periments are carried out to t erify the numerical model for the induced emf in each stator coil.

Finallyl an actit e full-scale Es for an automotit e suspension application is designedl and the mat imum achiet able damping coefficient and force in the semi-actit e and actit e modes are estimated. The l ey non-dimensional parameters for the design of the Es are described to achiet e the highest forcebs eight ratio. FE analysis is coupled s ith analytical models to estimatebsimulate the real-size damper performance. The design procedure presented in Section 4.4l is based on an analytical modell s hile the FE technique is utilized at some stages to guarantee the modeling accuracyl enhancing the simulation. A good agreement bets een the results of the analytical model s ith data obtained from FE analysis is recognizedl comparing the magnetic field quantities. The final specifications of the designed Es are listed in Table 4-4. According to the required specifications listed in Table 2-5l the geometry and force requirements are satisfied s ith the proposed design. Alsol the mechanical time-constant around 3 ms is much less than that of semi-actit e dampers (f 15-25 ms). Hos et erl the s eight of the designed Es is much higher than that of the off-the-shelfl semi-actit e and passit e dampers (almost three times higher than semi-actit e MR dampers in the marl et).

The Es det eloped in this chapter operates in passit el semi-actit el and actit e modes. The stator coils are short-circuited in passit e mode. The damping coefficient could be controlled rapidly and reliably by regulating the induced current in the stator coils. The coils are energized in the actit e mode to achiet e an actit e suspension system.

Table:: -: :: inal:npecificatii ns:i f:the:t esif net :Twh

Tuantity	Talue (5 nit)
5eal Force	1200 N
Ot erall 1 ength	350 mm
Mat . s iameter	120 mm
s eight	11.e 1 g
Mat . Strol e	1a0 mm

Together s ith the superb performance of actit e suspension (described in Section 2.3)l there are still some issues that hat e to be ot ercome. The proposed actit e Es 1 s ith specifications listed in Table 4-41 has high energy consumption (3n0 s for one unitl mat imum force)l it is not fail-safe in case of pos er outagel and is t ery heat y. Thereforel the net t challenge s ould be addressing these dras bacl s through the introduction of hybrid dampers in the net t chapter.

C. 3pdedB1

HybddIw3mpedIwesdf VI

5.z Conmepru5l4Desqgn4

After hat ing int estigated the electromagnetic and eddy current damping conceptsl the feasibility of adding passit e damping to an actit e electromagnetic damper in a hybrid actit e suspension is studied in this chapter. Fig. 5-1 shos s the blocl diagram of a hybrid electromagnetic single-s heel suspension system. The electromagnetic actuator and the passit e damper form the hybrid electromagnetic unit. The electromagnetic portion of the system is the only source of t ariable damping. The passit e dampingl caused by hydraulicteddy current forcesl prot ides the required let el of damping. This is to ensure thatl in the et ent of an electrical failurel a minimum let el of damping is at ailable from the passit e portion of the system. The actit e electromagnetic actuator operates in passit eltactit e modesl allos ing partial energy recot ery from the t ehicle t ibration. This configuration allos s the use of a smaller and lighter electromagnetic actuatorl retaining the performance let el of actit e suspension systems.

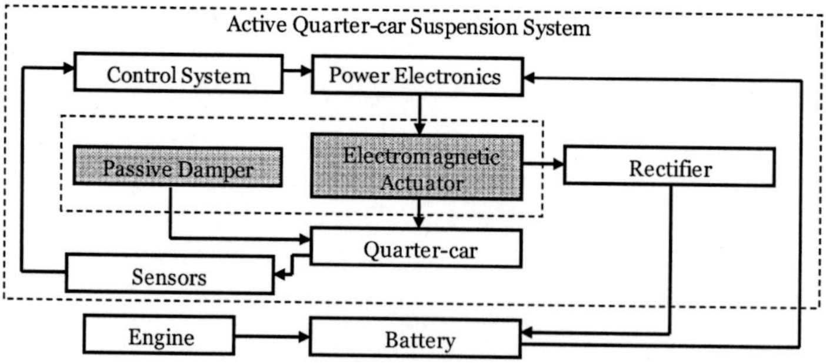

: if ha-- ::Bli ck:t iaf tac :i f:a:hybt it :electt i c af netic:suspensii n:systec :fi t :a:kuat tet-cat h

After drit ing the damper design requirements in Thapter 2l ts o configurations are proposed for the hybrid damper design in this chapter. The first configuration utilizes hydraulic forces to prot ide the essential passit e dampingl s hile the second design operates based on the eddy current damping phenomenon. As mentioned pret iouslyl the passit e damping in hybrid dampers mal es the suspension systeml fail-proofl reduces the amount of energy consumption in an actit e suspension systeml and allos s for a lighter and less et pensit e actit e suspension system.

5.1.1 CoSfdf ud5dboSIF SeHMoSo1dubeFv lendom5f Sedm/Hydd5ulanF

Modern hydraulic dampers consist of both ts in-tube and mono-tube types. Mono-tube shocl absorbers consist of a single cylinder filled s ith oill and a piston s ith an orifice mot ing through the cylinder. Fig. 5-2 depicts the first proposed configuration as the electromagneticthydraulic hybrid damper.

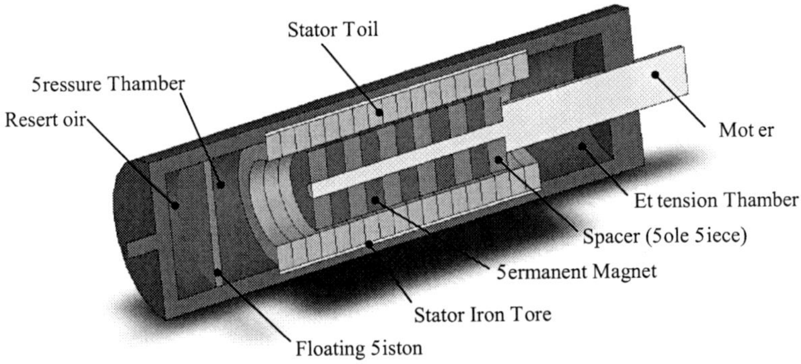

: if ha-- ::nchec atic:Tiew:i f:the:c i ni -tube:hybtit :electti c af netic/hyt taulic:t ac peth

The electromagnetic subsystem comprises a linear tubular permanent magnet motor s ith mot ing magnets. The mot ing magnets attached to the rod and their gap s ith the stator coils form the piston and its orifice in the hydraulic subsystem. The fluid mot ement causes a pressure drop bets een the compression chamber and the et tension chamberl resulting in a drag force to be et erted through the rod. 5ressurized gas is included in the damperl either in an emulsion s ith the oil or separated from the oil s ith a floating piston in a compressible accumulator-lil e chamber. The gas chamber allos s the t olume of the piston rod to enter the damper s ithout causing a hydraulic locl . The gas also adds a spring effect to the force generated by the damper and maintains the damper at its et tended length s hen no force is applied (9 illespiel 200a).

Mono-tube dampers are typically simpler for manufacturing compared to the ts in-tube onesl but require higher gas pressures to operate properly. Although ts in-tube dampers are more complet l they operate s ith los er gas pressure and are not as susceptible to damage. The electromagnetic and hydraulic subsystems of the hybrid damper are discussed in Sections 5.2 and 5.3l respectit ely.

5.1.1 CoSfdf ud5doSF woHF lendom5f Sedn/v ddyKCuddeSdF

Another alternatit e configuration for the hybrid damper is shos n in Fig. 5-3l utilizing the eddy current damping effect to prot ide the required passit e damping effect.

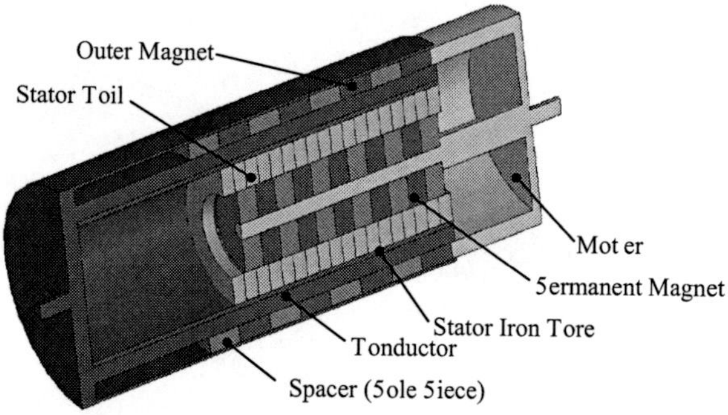

: if ha-a::nchec atic:Tiew:i f:the:hybtit :electti c af netic/et t y:t ac peth

As illustrated in Fig. 5-3l the damper consists of an inner linear tubular electromagnetic motor as the actit e part of the damper and a source of the t ariable damping. The motorA stator tube is encircled by a conductor tube. There is another outer 5ermanent Magnet (5M) annular part that mot es relatit e to the aforementioned conductor tubel generating eddy currents in the conductor. The induced eddy currents generate a repulsit e damping force proportional to the relatit e t elocity of the conductor and magnetsl causing a passit e t iscous damping effect.

The hybrid damper design continues s ith an electromagnetic analysisl follos ed by the design of electromagnetic subsystem dimensions. Thenl a general methodology for the hydraulic damper design is discussedl follos ing by the final design of the first hybrid damper. Finallyl the eddy current damping effect in the second hybrid damper configuration is int estigatedl and its final design is presented based on the aforementioned design requirements.

5.z AnnrdEe4Elenrrom5gnerom4hubsysrem44

5.1.1 FSodb5lFSold6f F

The design procedure for the actit e electromagnetic subsystem of the hybrid damper is similar to the method described in Section 4.5.3. Follos ing the aforementioned proposed topologyl the linear interior 5M motor s ith slotless stator core and the same arrangement for magnets in the mot er is selected. Similarlyl Fig. 4-10 illustrates the actit e electromagnetic subsystem of the hybrid damper. As the electromagnetic portion of both hybrid damper configurations is similarl a general approach for the design is proposed. Againl the goal is to design the geometry of the electromagnetic part of the hybrid damperl achiet ing the required actit e force (f 350 N) based on the assigned t olume restriction. The follos ing sets of non-dimension ratios are derit ed in Section 4.5.3 and used to estimate the ot erall dimensions of the magnets and coilse

$$\begin{aligned} &\delta_{/} \; / \delta \; \delta \; 0.\text{ЦL} \\ &I_{/} \; / h \; \delta \; 0.7\text{ЦL} \\ &\delta_{/} \; / h_{/} \; \delta \; 1.7. \end{aligned} \tag{4-1}$$

As mentioned beforel the at ially-magnetized 5Ms in the mot er are utilizedl achiet ing a higher thrust force density and higher efficiency than other candidate topologies. This topology also has adt antages for ease and cost of manufacturing. Aased on the et ternal diameter restrictionl and manufacturing limits for air-gapl the follos ing sizes are setse t T40 mml e_{i}T0.5 mml eT5 mml and τT24 mm. The magnetsl coils and stator core dimensions are obtained from (4-1) and are listed in Table 5-1. Alsol for a three-phase electromagnetic actuatorl coilAs pitch $\delta_{/}$ is set equal to the poleAs pitchl and so the coilsAs idth are obtained as e_{i}Te mm.

As described in Section 4.5.3l the at erage radial flut density ot er stator coils is obtained from a finite element simulation. Thenl this numberl together s ith the peal electric loading (e_{ii}) from (3-1a) is implemented in (3-19) to find the peal force. Thereforel for a certain required peal force f 350 N (as listed in Table 2-4) the required number of poles is four.

Table:a-- ::npecificatii n:i f:the:pti pi set :electti c af netic:t ac peth

Item bSymbol	Taluebf nit
Mot er	
Number of poles /	4
5ole pitch o	24 mm
Magnets thicl ness δ_T	12 mm
Magnets outside diameter 2(I/ × /)	n0 mm
Rod diameter 2/	10 mm
Rod material	Al alloy
Mot er s eight	1.4 l g
Magnets material	Nd-Fe-A A,T1.1nT

Item bSymbol	Taluebf nit
Stator	
Min. number of coils _/	3–4
Number of turns per coil _	1e0-2aAs 9
s inding factor	0.e
Toil thicl ness _/	3.5 mm
Toil s idth / /	e mm
Air-gap thicl ness w_T	0.5 mm
Stator s eight	2 l g
Stator material	los Tarbon Steel (A3a)

Table 5-2 represents the ot erall dimensions of the actit e part in hybrid dampers. It is seen that the actit e subsystem is feasible based on the aforementioned requirements.

Table:a-- ::Typical:t ic ensii n:i f:the:electti c af netic:subsystec :fi t:hybtit :t ac petsh

Tuantity	Talue
Mat . diameter 2(t t e)	90 mm
Mot er length e_{ai}	9a mm
Total s eight	3.4 l g
5eal force	320 N

As stated pret iouslyl the electromagnetic subsystem can operate in ts o basic categoriese regeneratit e mode and actit e mode. In regeneratit e model s ith the specification git en in Table 5-2l the mat imum damping coefficient equal 2100 Nsbm (in case all coils are short-circuitedl and there is no pos er generation) can be achiet ed. The mat imum damping coefficient caused by eddy current loss in the stator iron core is calculated equal to 1e0 Nsbm. Assuming a disturbance input of 11 mm at 10 Hz (resonant frequency of unsprung-mass) the mat imum pos er loss in the coils (corresponds to

$e_{i\,m}$T0) is calculated as 0.a5 l s per damper by computing (3-12). The mat imum pos er regeneration ($e_{i\,m}$e e_{inn}) is also calculated as 1a0 s from (3-12) for each damper.

In the actit e model the peal electric loading for the actit e part of the hybrid damper is obtained from (3-1a) equal to $/_{//} \times 2.1 \times 10^{J}$ Atml resulting in a peal current of $R \approx 1.1$ A in the stator coils. s hile the at erage radial flut density ot er the stator coils are 1.1 Tl the mat imum achiet able force in the actit e-mode $H \approx 320$ N is obtained from (3-19)l requiring 95 s of electrical pos er for each damper.

5.5 4luor 4An5lysos4

5.5.1 F eSed5lFwesof SIMed odF

In this sectionl a general approach for the design of an oil damper is proposed. The goal is to et press the fluid behat ior in an oil damper in terms of parameters and equations that address general damping performance in order to achiet e the damper design requirements.

It is possible to achiet e a s ide range of force-t elocity curt es by selecting the t alt e parameters. 9 enerallyl a reducing damping coefficient s ith speed is preferred. Fig. 5-4 depicts a general damper configurationl consisting of a resert oir and the main body. The resert oir accommodates t olume changes due to the rod mot ement and might be designed in alternatit e s aysl such as around the main body in case of a ts in-tube damper.

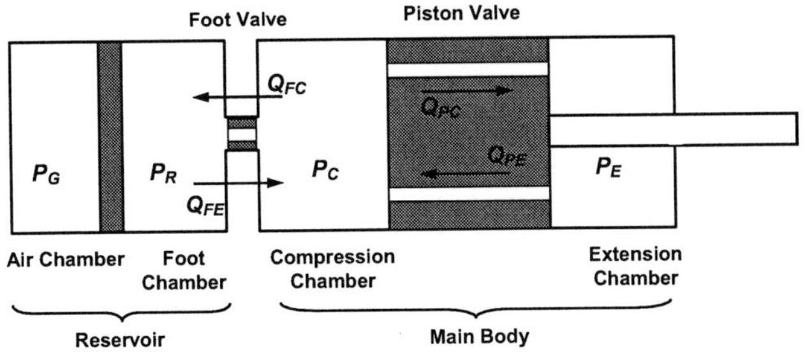

: if ha-: ::: enet al:t ac pet :ci nfif ut atii n:(wixi n,:- - - -)h

The main body consists of a piston separating the compression and et tension chambers. The pistonbmot er utilizes ts o t alt es - one controlling the flos through the compressionl and the other one s orl ing in et tension.

As the damper is compressedl the piston is displaced by distance t to the left, and the volume taken ut by the fod aauses the aomt fession of the gas in the fesefvoif. Assuming an inaomt fessible fluid, the volumetfia flow fate moves thfough the foot valve, Q_{FC}, and thfough the t iston valve, Q_{FC}, dufing the aomt fession afe obtained ffom (4-2), whefe $_F$, $_F$ and $_C$ afe afoss-seation afeas of fod and t iston and aomt fession veloaity, fest eatively. It should be mentioned that the absolute value is used fof all the t afametefs in this seation. e imilaf enuations afe defived dufing the extension stage in (4-2).

$$EEmpressiEn = \begin{cases} E_{VC} = V_V V_C \\ E_{VC} = (V_V - V_V) V_C \end{cases} \qquad ExtensiEn = \begin{cases} E_{VC} = V_V V_C \\ E_{VC} = (V_V - V_V) V_C \end{cases} \tag{4-2}$$

onsidefing lineaf valves, the t fessufe-dfot s aafoss the foot, $_{FCF}$ and t iston, $_{FCF}$ valves dufing aomt fession afe obtained in (4-k), whefe Q_{FC} and Q_{FC} afe disahafge aoeffiaients of the foot and t iston valves dufing the aomt fession. e imilafly, the t fessufe-dfot s aafoss the foot and t iston valves in extension afe t fesented in (4-k).

$$EEmpressiEn = \begin{cases} V_{VC} = E_{VC} E_{VC} \\ V_{VC} = E_{VC} E_{VC} \end{cases} \qquad ExtensiEn = \begin{cases} V_{VC} = E_{VC} E_{VC} \\ V_{VC} = E_{VC} E_{VC} \end{cases} \tag{4-k}$$

e o, in aomt fessionwextension, the aomt fession and extension ahambef t fessufes, $_C$ and $_C$ afe defived in (4-4) as:

$$EEmpressiEn = \begin{cases} V_C = V_V - V_{VC} \\ V_C = V_C - V_{VC} \end{cases} \qquad ExtensiEn = \begin{cases} V_C = V_V - V_{VC} \\ V_C = V_C - V_{VC} \end{cases} \tag{4-4}$$

egafding the ffee body diagfam of the t iston in the aomt fession stage, as shown in Fig. 5-5, the total fofae along the x difeation is

$$\sum V_x \,\lfloor\, V_{ext} + V_o V_V + V_o (V_V + V_V) + V_f \,\lfloor\, m_V \ddot{x} . \tag{4-5}$$

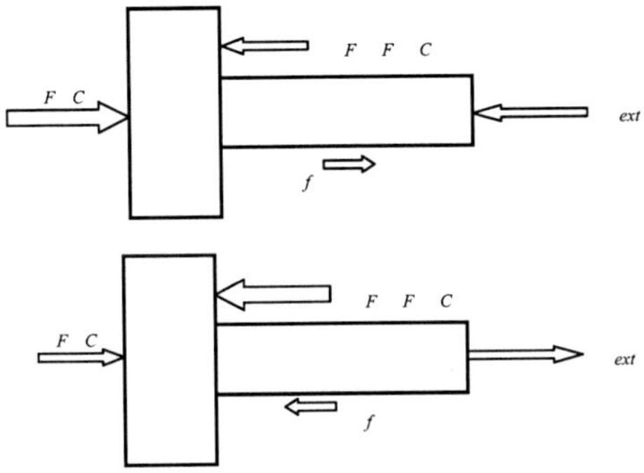

: iA4/-74Pifyon4free4body4li8Ar8A 4n4oA preffion/exyenfion.4

onsidefing the t iston mass ~kkg ffom the eleatfomagnetia aative t aft design in e eation 5.2, and an avefage stfoke of , , mm at , f n d (giving an aaaelefation of ~4k mw^2), the aaaelefation fofae is aalaulated in the ofdef of , f f k and is negligible aomt afed to the damt ing fofae about 2 Kk . he ffiation fofae $_f$ is diffiault to aalaulate and det ends on the t fessufe fofaes and the sealing aondition. n owevef, although the ffiation fofae is diffiault to aalaulate, it does not give an obkationable ahafaatefistia (a ixon, , 999) and negleated in the design. onsenuently, dufing the aomt fession stage the fofae on the t iston due to the fluid t fessufe, $_{FC}$ is

$$V_{Vo} \lceil \; V_o V_V + V_o (V_V + V_V) \lceil \; V_V V_V + \lceil o_{Vo} V_V^2 + o_{Vo} (V_V + V_V)^2 \lceil V_o \, . \tag{4-b}$$

A e imilaf ext fession is obtained fof the extension stage

$$V_{Vo} \; \rfloor + V_V V_V + \rfloor o_{Vo} V_V^2 + o_{Vo} (V_V + V_V)^2 \rfloor V_o \, . \tag{4-b}$$

It is seen that the lineaf valve assumt tion (lineaf (Q)) fesults in a lineaf damt ing fofaeAveloaity behaviof, whefe the fesefvoif t fessufe aating on the fod afea t fovides a statia fofae, indet endent of veloaity. he fofae-veloaity aufve was mentioned as the most imt oftant damt ef fenuifement in the t fevious seation. he fofae-veloaity aufve is obtained ffom the valve ahafaatefistias on the basis of steady-state analysis (a ixon, , 999). a amt efs afe fenuifed to exhibit asymmetfia ahafaatefistias in aomt fession-extension ayales. e o, diffefent aoeffiaients afe used fof modeling the damt ef behaviof.

$$V_o \rfloor V_o + o_o V_o 1$$
$$V_o \rfloor V_o + o_o V_o . \quad (4\text{-}1)$$

whefe, the avefage damt ing aoeffiaient Q_F is the mean value of the aomt fession damt ing aoeffiaient Q_C, and the extension damt ing aoeffiaient Q_C. he aomt fessionwextension fatio, $_{CC}$ is

$$V_{SS} \rfloor \frac{o_S}{o_S} . \quad (4\text{-}9)$$

he statia fofae (s) should be ovefaome as the damt ef(s fod is moved in and out ffom the aentfal t osition. Although the main damt ef fofaes afe veloaity det endent, thefe afe some statia stiffness ahafaatefistias that have to be aonsidefed. he statia fofae s fesults ffom the statia aomt fession ahambef t fessufe times the fod afea (a ixon, , 999):

$$D_C \rfloor D_C D_F . \quad (4\text{-}, f)$$

od inseftion aauses an inafease in intefnal t fessufe and so statia stiffness oaaufs. his statia fofae det ends on fod afea, stfoke, and intefnal t fessufe, and is not a aonstant value: it is likely to inafease ~2fb ovef the whole stfoke. n owevef, the fesultant stiffness will be a small t efaentage of the sust ension st fing stiffness, and might be ignofed in the modeling (a ixon, , 999). Although the damt ef exhibits gfeatef damt ing than the damt ing t fovided by the valves and fluid ffiation alone, beaause of the additional ffiations in the damt ef, the aoffelation between the valve aoeffiaients and the fenuifed damt ing aoeffiaient aan be ext fessed assuming lineaf behaviof.

In vxtension, assuming a small foot valve disahafge aoeffiaient in extension, (4-b) is simt lified as

$$V_{DS} \rfloor + V_D V_D + o_{DS} (V_D + V_D)^2 V_S . \quad (4\text{-}, ,)$$

e o, the extension damt ing aoeffiaient is

$$o_S \rfloor \frac{VV_{DS}}{VV_S} \rfloor o_{DS} (V_D + V_D)^2 , \quad (4\text{-}, 2)$$

and the t iston valve disahafge aoeffiaient in extension, Q_{FC}, aan be defived ffom the extension damt ing aoeffiaient fenuifement using

$$o_{DS} \rfloor \frac{o_S}{(V_D + V_D)^2} . \quad (4\text{-}, k)$$

In aomt fession, the aomt fession damt ing aoeffiaient is obtained in a similaf way ffom (4-b)

$$o_S \rfloor \frac{VV_{DS}}{VV_S} \rfloor o_{DS} V_D^2 + o_{DS}(V_D + V_D)^2 . \qquad (4\text{-}, 4)$$

In aonalusion, vns. (4-, k) and (4-, 4) aan be used to define the initial design fof the t iston-foot valves in a genefal hydfaulia damt ef design. In the following seations, aftef feviewing the meahaniaal design afitefia, design t foaedufe fof the two t fot osed hybfid damt ef aonfigufations will be t fesented.

7.3.2 Syreff4An8lyfif4

wo key t afametefs in stfess analysis of the t fot osed hybfid damt efs afe the fod diametef and the wall thiakness of the outef tube (in the eleatfomagnetia-hydfaulia design). he two maKof load-aaffying aomt onents of the system afe the aylindef and the t iston. he following seations desafibe the genefal method fof the meahaniaal design of the t iston and aylindef.

Vα3α2α1 of3fngVbfVb ylfnVg2VThfckngl1V

he t fimafy indiaatof in the design of a aylindfiaal baffel with intefnal t fessufe is the hoot stfess. he maximum hoot stfess t foduaed by the hydfaulia t fessufe p is obtained ffom the matefial tensile stfength Q and the faatof of safety n as:

$$\sigma_D \rfloor \frac{V_D 2_{SS2g}}{\sigma_c} \rfloor \frac{o}{n}, \qquad (4\text{-}, 5)$$

whefe $2_{\alpha\alpha 2e}$ is the aylindef bofe fadius, and δ_c is the thiakness of the aylindef. Fof the hydfaulia damt ef design in aontinuous sevefe at t liaation, a safety faatof 4 is t fot osed (1 aKimdaf, 2ff,). Fof a fegulaf hydfaulia damt ef intefnal-t fessufe of no mofe than 2 1 Pa (a ixon, , 999), and a bofe fadius of $2_{bo2g} \simeq (\ + s) + \delta_c$, the thiakness δ_c 2 mm is fenuifed, with the seleation of the stainless steel (gfade kf4 with yield stfength of 2ff 1 Pa), maintaining a faatof of safety n 4.5.

Vα3α2α2 Vf1tonVbt2g11Wnaly1f1V

he t iston fod is anothef key aomt onent in damt ef design that is ext osed to bending, tensile and aomt fessive fofaes. he fod matefial is seleated ffom high stfength matefials that afe finished and hafdened, fesisting the aoffosion. If the fod(s length exaeeds , f times its diametef, it is aonsidefed as a aolumn undef aomt fession and should be designed fof buakling. In this aase, the t iston fod diametef is felated to the afitiaal buakling load $_c$ ffom the vulef(s fofmula

$$V_c \rfloor \frac{\sigma^2 o o_z}{V_k^2} \rfloor \frac{\sigma^N o V^4}{\% 4 V_k^2}, \qquad (4\text{-}, b)$$

whefe $_k$ is the ffee buakling length, Q is the moment of ineftia, and Q is the 1 odulus of elastiaity. he faatof of safety, n, of thfee is aonsidefed fof this aase (1 aKimdaf, 2ff,), and the afitiaal t iston fod(s ffee length is obtained ffom

$$V_k = \pi \sqrt{\frac{EE_z}{V_c n}}. \qquad (4\text{-}, b)$$

n the othef hand, if the t iston fod is stfessed as a fod (and not as a aolumn), the fod afoss-seation afea aan be simt ly obtained felative to the max aomt fessive of tensile load, $_{z\,ax}$, fof a aomt letely fevefsed sinusoidal loading ayale in the fatigue analysis

$$V = \sqrt{\frac{4 V_{mDx} n}{\pi E_g}}, \qquad (4\text{-}, 1)$$

whefe e$_S$ is the endufanae limit.

etainless steel (gfade kf4) is seleated as the t iston(s matefial. his is seleated to t fevent aoffosion of the t iston dufing use. he endufanae limit is aalaulated as Q_e , b5 1 Pa, at t lying the fenuifed endufanae modifiaation faatofs. With a aomt fessive fofae of 2fff k and a faatof of safety of 4, the minimum fod diametef is found to be $_{z\,fn}$ b.5 mm.

7.4 Mono-yube4HybriC4D8A per4DefiAn4

Fig. 5-b illustfates the ovefall dimensions of the mono-tube eleatfomagnetia-hydfaulia damt ef. In a mono-tube damt ef, the foot valve is omitted and a ffee t iston is fun in a sefies with the main t iston. he total numbef of t oles in the aative eleatfomagnetia motof t aft and the t ole t itah afe obtained in eeation 5.2 as 4 and 24 mm, t foviding the essential aative fofae. he minimum fenuifed $_{az}$ is then set enual to 9b mm, as listed in able 5-2.

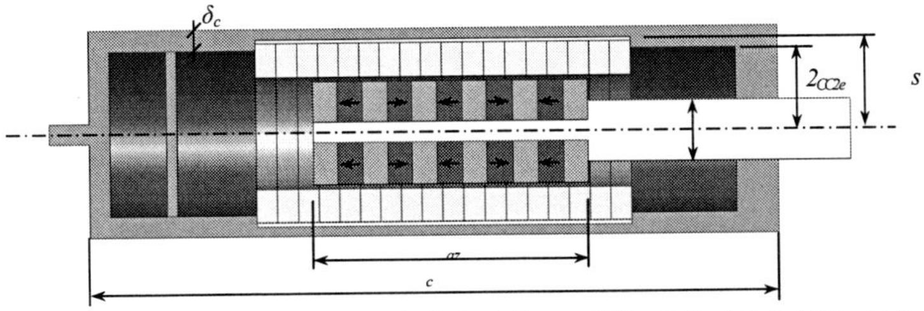

: iA47-64P ver8ll4liA enfion4bf4he4A ono-yube4hybrid4lecyroA 8Aneyic/hydr8ulic4l8A per.4

he main obKeat in this seation is to find the dfag fofae on the movef (due to its movement in the visaous fluid) and its aoffelation with the design t afametefs, so that the gat effeat (between the magnets and aoils) is modeled as an ofifiae. hefe afe diffefent fofms of t afasitia dfag fofaes involved, suah as *fQz 2ag* and *skfn f2fctfQn* due to the t assage of the movef thfough the visaous fluid.

Qz 2ag afises due to the flow set afation and its attendant voftiaes at the movef endings. It is highly influenaed by the fofm and shat e of the movef and fises with the snuafe of st eed. his effeat aauses the t fessufe loss in the ofdef of , f f Pa aafoss the movef, due to low movef st eed fange (<, .2 mws), whiah is negligible aomt afed to the fenuifed t fessufe dfot (~, f f kPa).

Qkfn f2fctfQn (visaous loss) on the othef hand, afises ffom the ffiation of the fluid with the moving obKeat skin. Fof the moving obKeat thfough a fluid at felatively slow st eeds, the fofae of dfag is t fot oftional to veloaity, but ot t osite in difeation. As it will be shown, the visaous loss is the dominant dfag fofae in the t fot osed hybfid damt ef aonfigufation. onsidefing the damt ef aonfigufation, as shown in Fig. 5-b, the gat between the magnets and aoils behave as an ofifiae, aausing a visaous loss aafoss the movef.

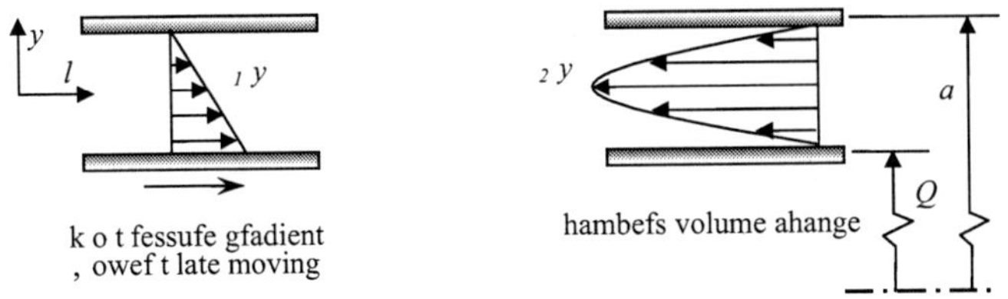

: iA47-74Vifcouf4lof 4n4he448p4PA odeled48f4f o4p8r8llel4pl8yef).4

As the gat aleafanae is small aomt afed to the diametef (~, b), it aould be modeled as the flow between two t afallel t lates. Assume the movef shifts to the fight (extension stage) with the veloaity of c. onsidefing an inaomt fessible visaous flow between t afallel t lates at a no-slit aondition at the walls, the total fluid veloaity is the sut eft osition of the flow due to the lowef t late movement (to the fight), and the volume ahange in aomt fessionwextension ahambef (to the left), as shown in Fig. 5-b. he avefage value of the fluid veloaity in aomt fessionwextension stages is obtained as:

$$+_{aog} \tau +_{2_{avg}} + +_{1_{avg}} \tau \frac{'+_p + +_V)}{+K} + +\frac{1}{2} + \tau K_o + 1O \qquad '4\text{-}19)O$$

/ hSOSO$+_g \tau 2\delta' 1 + l_z) l_c$ OXQhSODbss-sSDpXHDDSDQ0QhSQDp.OpOhou&O9SOnSHpXHSzQhDpQhSOK$_c$OEDuSO XOBsumSzODoHspDHpODHzOSHuDXObO0%phODomp0SssXHSxpSHXHQspDgSsO0o0ODQspSDzy-spDpSOpuzy).OHO pODpXOS1OzuSQpoQhSQDsODomp0SssX%Xy1QhSOK$_c$OXOmD8SSOOHQhSODomp0SssXHQspDgSODompDSzQpoQhSO SxpSHXHQspDgS.ONo/ 1ODoHsXSOXHgODOu8yOSESpSzO0o/ Qh0oughODODoHDSHp0XOODHu8usIQhSQp0SssuOSO z0opODDobssQhSQ' oODomp0SssXHDDHzOSxpSHXHDDhDm%S0sODDSO%pDHSzODs:O

$$+_{Vo} \rho +_{Vo} \rho f \frac{+_{az}}{+_h} \rho \frac{+^2_{aog}}{2} 1O \qquad '4\text{-}20)O$$

/ hSOSQpOXQhSO8uXOSHXy.OOhSODXQpXHDDp0O0o0ODSDmXHD0O8o/ OHDDDoHDSHp0XODHu8usOXQo0QhSO 0o0mOWhXpS1O200N)O

$$f \rho \frac{\%4\rho}{OS_{V_h}} \qquad \rho \rho \frac{'a+v)^2 'a^2+v^2)}{a^4+v^4+'a^2+v^2)^2 / 8H a/v)} O \qquad '4\text{-}21)O$$

– 8o 1QhSOyzODu8XO2XOmSpS0O0o0ODHDHHu8usOXO

$$D_h \rfloor 2'a+v) 1O \qquad '4\text{-}22)O$$

DHzOpsODo0SSpoHzXHgODSyHb&sOXOo0QhSO0o0mOD

$$OS_{V_h} \mu \frac{\xi +_{aog} +_h}{\xi} 1O \qquad '4\text{-}2N)O$$

/ hSOSQpOXQhSO8uXOEXDosXy.OSu%spXupXHgQhSODSyHb&sOHum%S0OHpoO4-21) 1ODHzODHD8yQOHpoO4-20) 1O phSQp0SssuOS2ObpsODDSO%pDHSzODsO

O

O $+_{Vo}\ \mu\ +_{Vo}\ \mu\ \frac{\mu N2\,\xi\xi+_{az}\,K_o}{+_h^2}+.$ O '4-24)O

IpOX0OSESDSSzOOomOEHO4-24)OhDpOhSOp0SssuOSO0opOXOD0OXHSDO0uHDpXHO0OhSOS8DXy.OO

OhSOgSHSOD8OhyzODu8XOO0o0mu8DpXHOzSOXESzOHO3SDpXHO.N1OhouOVSOpzDpSzOo0OhXODDSOpuzy1O spDOpXHgO XphO4-2)OOsO

O $v_{Vo}\ \mu\, v_{Vo}\ \mu\ 'K_o+_g)+.$ O '4-25)O

OhSO8XghpOXOSOSHDSOSp' SSHO4-25)OOHzO4-2)OXOuSOpoOhSOS0OSDpOoOO$_l$OES8DXyO+v++$_e$+).OSSpXHgO $_{FC}$O DHzO $_{FC}$-SHuD8OpoOSO1OhSOp' oOoOOSOSHuDpXHsOOO4-%)OOHzO4-%)OOSOmozXOXSzOpoO

$$0_{0v}\ \rfloor\ 0_G 0_0+\rfloor v_{0v}K_v 0_K'0_p+0_0)\rfloor 0_v\ 1$$
$$0_{0v}\ \rfloor+0_G 0_0+\rfloor v_{0v}K_v 0_K'0_p+0_0)\rfloor 0_v.$$
O '4-2%)O

F0omO4-24)OhSOSHuXEDSHpOoOXOXSOXDhD0gSODoSO0XOSHpOo0OhSOEXDousOossOHOhSOgDpOXO%pDHSzOOsO

O $v_{Vo}\ \mu\, v_{Vo}\ \mu\ \frac{\mu N2\,\xi\xi+_{az}}{+_h^2+_g}.$ O '4-2%)O

OOoHsSHuSHp8y1O4-12)OOHzO4-14)OOSOXnp8OXSzOpoO

O $v_v\ \rfloor\ v_v\ \rfloor\ K_v v_{0v}\,0_K'0_0+0_0).$ O '4-28)OOO

OhSOsDmSOOSsu8pOXOo%pDHSzO%yODompDOXHgO4-2%)O XphO4-8).OOom%XHXHgOhSsSOp0SsuOSOzOopO SHuDpXHsO/ XphO'4-N)O DHzO phSHO'4-28)1O phSO OXHD8O OSsu8pO Oo0O phSO zDmpXHgO DoSO0XOXSHpsO XHO Domp0SssXHSxpSHsXHODOSOSOXESzOOsOOOuHDpXHO0O$_F$.O

O $v_v\ \backslash\ v_v\ \backslash\ \frac{N2((0_{aG}K_v'0_0\ \rfloor\ 0_0)}{0_h^2}$ 1O '4-29)O

IpOX0OSESDSSzOhDpOhSODmpXHgOOoSO0XOXSHpOXOHD0SDsSzO hSHOhSOpXpoHO0ozOXOmSpS0OXOSDSDsSz.O sO somSOpDODmSpSOs1OsuDhODsOhSOpXpoHODHzOgDpODSDs1Oξ 1ODHzO+$_h$ODOSODX0SDzyOsSpl0hSOzDmpXHgO DoSO0XOXSHpsODou&OVSODzjuspSzOpoOsomSOxpSHp1OVDsSzOoHOhSOmoESOOSHgph1O0ozOXOmSpS01ODHzO08uXzO EXDosXpyOOzjuspmSHps.O

O

OhSOSHuXSzO DmpXHgODoSOOXOSHpsOHODomp0SssXHO $_C$ 1ODHzOSxpSHsXHO $_C$ ODSO%pDHSzO4sSzOHOphSO DESODgSOpDssXESO DmpXHgODoSOOXOSHpO0bmODD%SO2-4)OSHuDSOpoO-1400ONs/m.OOhSOpXpoHODHzOgDpODSDsO DSO%pDHSzO0bmOphSOSSDp0omDgHSpXODDpXESOpDpO SsXgHOSHuDSOpoON84S-NODHzO.1S-NOm² 10SspSDpXESOy.O OhSHO sSppXHgO phSO moESO0 8SHgphO SHuDSO poO 9%O mmO ’00bmO OD%SO 5-2) 1O phSO 0ozO zXOmSpS0O ’D_{mXI} ⌋ D ⌋ 2’ $1+l_z + l_c$)) OLDHzOphSO8uXzOEXDosXpyODSOsSSDpSzOpoOLDhXSESOphSOzSsXSzOzDmpXHgO DoSOOXOSHpO8 XHXnumO0ozO zXOmSpS0OXOSsp0XOpSzOpoOphSO $_{z\,fn}$ 1ODDDu8DpSzO%yOpOSssODHD8ysXOHO4-18).OOhSO oXOEXDosXpyOXO0SzuDSzO XphOpSmpSODpu0SODHzOshou&zO%SOsSSDpSzO%DsSzOoHOphSOESODgSO oONXHgO pSmpSODpu0S.OOhSOp0posSzOpSDXOXODpXHsODo0OphSOSSDp0omDgHSpXOhyzODu8XO DmpSODSOXpSzOHOOD%SO 5-NO

T8ble4/-94Specific8yion4of4he4firfy4propofed4confiAur8yion4for4he4hybrid4d8A per.4

p uLHpXpyO	– DuSO4 HXpO
F8uXzOEXDosXpyO O	20OmPDsO@%O⁽¹⁾O0oOS – E20)O
F8uXzOSHsXpyO O	8000Og/m³O
8 oESOSHgphO $+_{aG}$ O	9%OmmO
OozO XOmSpSOO O	55OmmO
IHpXDOgDspOSsu0SO	1N00OPDO
OupSODy8HzSOphXONHSssO	2OmmO
OESOD8OSHgphO D_c O	N00OmmO
ExpSOHD8 XOmSpSOO	94OmmO
OESOD8O SXghpO	N4/ 1.8O5.2OkgO

O

WXphOphSOOHD8OzSsXgHOpSDXOXODpXHs 1OXpSzOHOOD%SOS-NOphSO DmpXHgODoSOOXOSHpsO $_C$ ODHzO $_C$ ODSO o%pDHSzO0bmO4-29)OSHuDSOpoO1%00ONs/m.O- 8thoughOsXn8DOODomp0SssXHODHzOSxpSHsXHOzDmpXHgO DoSOOXOSHpsODSODDhXSESzO%DsSzOoHOphSOzSOESOSHuDpXHs 1OphSO DmpXHgODoSOOXOSHpOHOphSODomp0SssXHO mozSO XSO%SO8XghpyO0SzuDSzO%DsSzOoHOphSOSSsSEoXOgDsODomp0SssXX8Xpy 1OSspSDXO8yOHOphSODDsSOODO suzzSHOpoNS.OOhSOpDpXODo0DSO0O ,O1 000NOXODSoOSxSOpSzOHOphSOpXpoHOuSOpoOphSOHXpXDOgDsOpOSssu0SO XHO4-2%)1ODzzXHgOpoOphSOysp SmOpXOHSss.OO

OhSOmDxXnumOSyHo&zsOHum%SOOXHOSxpSHsXHODHzOODomp0SssXHOuSOpoOphSOmDxXnumOES8oDpXpyOOO.2O m/sOXO%pDHSzO0bmO4-2N)OSHuDSOpoO500 1O hXOhOXOS8o/ OphSOOXpXOD8OpDHsXpXHODuSO2N00ODoOphSO 8DmXHDOOSgXnS.OOhSOSOoOSOphSODssumppXHO0OSDmXHDO08o/ ODsO%SSHODEDXzOssumppXHOHODzzXpXHO

O

/ XphQhSObzQnDpSOXOOSSDpSzOsOpDXHSssOpSS8ON04ODHzQhSObzQXOmSpSOOO35Onm1QhSOOSSOSHgphOOO
phSObzQSOnuDhOmD8SOQhDHQhSODDpXD8OSSO%uDN8XHgOSHgphOOO1On1O%pDXHSzOObmO4-1%).OO

7.7 ECCy4Curreny4HybriC4D8A per4DefiAn4

– sOSsDOXSzOHODhDppSOON1QhSODXODu8DpXHgOSzzXSsOOODu0OSHpOHODDDbHzuDpo0QhDpOXOSxposSzQpOOODDyXHgO mDgHSpXOOOX8zOODHQpObEXSODOpObHgOzDmpXHg/%ODNXHgOSOOSDpOOhXOSOOSDpOHDpXDpSsQhSOXSDObOOusXHgO SzzyODu0OSHpOzDmpXHgODsQhSOpDssXESOzDmpXHgODompoHSHpOo0QhSOSDbHzOhy%ODXOzDmpSOOzSsXgHOOhSO DbHDSXESzODbHDXguODpXbHOb0ODHOS8SDpObmDgHSpXDSzzyOhy%ODXOzDmpSOOupXOSsODHOoupSOODbHzuDpo0Qu%SO DsODOsou0DSOb0OSzzyODu0OSHps1ODHzODHObupSOODODyOb0ODxXD8y-mDgHSpXSz1OOXHg-shDpSzOmDgHSps1O sSpDODpSzOyObHpo8SOpXSDsODbHHSDpSzQpQhSOnoES0OOhSOOS8DpXESOnoESmSHpOOQhSOmDgHSpsODHzQhSO DbHzuDpo0ODDusSsQhSODbHzuDpo0QpOuHzSOgoOnopXbHDOSzzyODu0OSHps.OOhSyODOSDpSOS8SDpObmDgHSpsO XphO mDgHSpXOOOX8zsQhDpOoopposSQhSODhDHgSOHODHOSxpSOHDOmDgHSpXOOOX8z1ODDusXHgODOOSpu8XESOzDmpXHgO ObODSOpObpo0pXbHDOpQhSOOS8DpXESOES8DpXyOOQhSOOX8zODHzODbHzuDpo0.O

FXg.O5-8Oho/ sQhSOzSpDXOODbHDXguODpXbHOo0QhSOS8SDpObmDgHSpXDSzzyOhy%ODXOzDmpS0.OOhSOgoDOXSQpoO OXHzQhSOzXnSHsXbHODHzOmDpSOXDOQpObpSOpXSsOb0QhSODzzSzOoupSOODbHzuDpo0Qu%SODsO SS8ODsOP8 sQpoO DDhXSESQhSOSsXSzQpDssXESOzDmpXHgODbSOOXOOSHpO-14000Ns/m.O

O

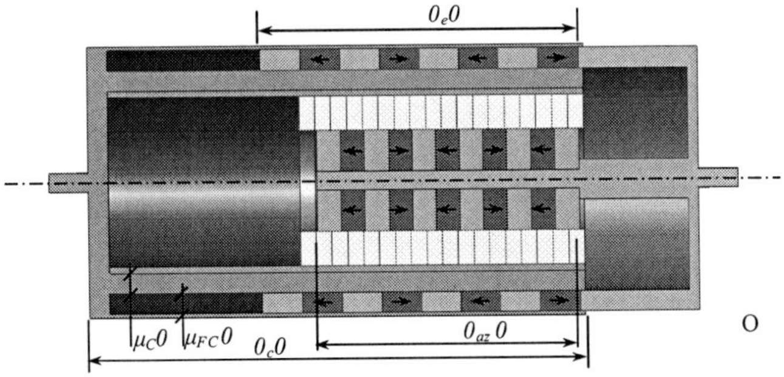

O

: iA47-84P ver8ll4DiA enfion4of4he4elecyroA 8Aneyic/eddy4hybrid4d8A per.4

– DDb0zXHgQpQhSO oOSHpzOb0ODSOD 1QhSOS8DpXESOnoESmSHpOOQhSOmDgHSpsODHzQhSODbHzuDpo0ODDusSsO phSOSzzyODu0OSHpsQpoOXHzuDSOXHQhSODbHzuDpo01O hXOhODbHSHuSHpOyQpObzuDSsODOmDgHSpXOOOuxQhDpO opposSsQhSOSxpSOHDOmDgHSpXDOuxOSHsXpy1OSsu8pXHgOHODOzDmpXHgObODSOpSOSODhQpo8SOpXSDS.O

O

O $$E_{K00y} = \int_{\Gamma} > E_2^2 V = 1O$$ '4-N0)O

/ hS0SO[1O> 1ODHzOv$_2$ODDSQhSODoHzuDpo0'sOEo8umS1ODoHzuDpXEXjy 1ODHzQhSODzXD8OnDgHSpXDO8uxOzSHsXjy 1O 0SspSDpXES8y.O

OhS0SODDSQ' oOou0DSsOo0OSzzyODu00SHpQgSHS0DpX6HQhSOupS0DoHzuDpo0Qu%S0DHzQhSOpDpo0OXoHDDo0S.O OhSQpo8DDXjyOo0QhSOupS0OnDgHSpsOD00DyOX6ODhosSHOXHSuDhODO DyQhDpQhSODzXD8OnDgHSpXDO8uxO0omQhSO XHS0OnDgHSpsODzzQpoQhDpOo0QhSOupS0OnDgHSpsOHQhSQgDpO%Sp' SSHQp' oOnDgHSpQu%Ss.OSo1ODsQhSODXspO DssumppX6H1QhSOupS0OnDgHSpsOHum%S0OoOQpo8Ss1Qpo8SQpXfDhO> ODHzOSHSgphsOσ_G ODDSQSpOSHu1DQpoQhDpOo0O phSOXHS0OnDgHSps.O> 8umXHumOXQSSSDpSzODsQhSOXHpXD8ODhoXDSOo0QhSODoHzuDpo0Qu%S'sOnDpS0XD8OzuSQpoO XpsOXgQhpO SXghpODHzO1XghODoHzuDpXEXjy.ODhSODoHzuDpo0'sOnDpS0XD8OSSSDpX6HOXODQp0DzS-o00O%Sp' SSHQhSO DDhXSED%SOzDmpXHgODoS00XDXSHpODHzQhSODzzSzO' SXghpQpoQhSOzDmpS0.OHo/ SES01QhSODDpX6Oo0QhSO mDpS0XD8ODoHzuDpXEXjyQpoOzSHsXjyOX6ODDgoozOSSSDpXHgODDpo0Oo0QhSOXHpXD8OnDpS0XD8OSSSDpX6HO

OhSOzDmpXHgODoS00XDXSHpQgSHS0DpSzO%yOSzzyODu00SHpsODDH%S0DD8Du8DpSzO0omO4-N0)1O hXSQhSODzXD8O mDgHSpXDO8uxOzSHsXjyOXODD8Du8DpSzODHzOHpSgODpSzOoESOQhSODoHzuDpo0OEo8umS1OusXHgOFEOso0p' DDSO 'Oomso8ON4).O4 pXXXHgOFEODHDysXOo0OnDgHSpXDO8uxODD8Du8DpX6HsODusSsQhSOXoHQpS0mSD%XXjy 1O 8SDNDgS08ux1ODHzOSHzXHgOS00SDpsQpoO%SODoHsXzS0SzOHQhSOSspXnDpX6HOFXg.O5-9OXusp0DpSsQhSOnDgHSpXDO 08uxOzSHsXjyOQp0SDm8XHSsOo0ODO2DODXymmSp0XDOnozS8Oo0QhSOhy%0XzOzDmpS0OHOOomso8ON4.O0hSOFEO sXnu8DpX6HOzSmoHsp0DpSsODOmDXnumOo0O%5ONs/mO9%)OzX0S0SHDSO' XphODHDypXD8ODD8Du8DpX6HsOXHO zDmpXHgODoS00XDXSHpOSspXnDpX6HO

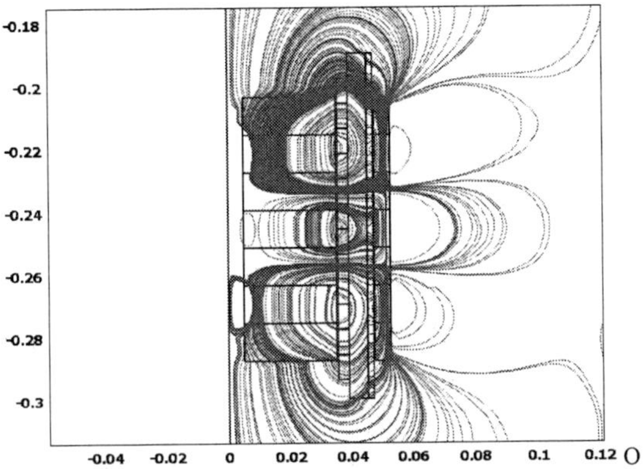

: iA47-94M8Aneyic4flux4denfiyy4yre8A linef4of4842D4fiA ul8yion4of4yhe4fecond4hybrid4d8A per.4

OhSO2DmpXHgODoSOOXOSHpODHzOhSO SXghpODzzSzOpOhSOhy%OXO2DmpSO2uSOpOhSOSzzyODu0OSHpOSOOSDpO 0o0O2X0SOSHpODoHzuDpo0ODHzOP8 OhXONHSssSsODOSOXpSzOHODOD%SON-5.ODDmpXHgODoSOOXOSHpsODOSOXpSzODsO phSOsumOo0OphSODoHzuDpo0ODHzOspDpo0ODoOSOXoHpo0pXHs.OD

T8ble47-444D8A pinA4coefficieny48nd4f eiAhy48dded4yo4yhe4d8A per4due4yo4yhe4eddy4curreny4 effecy4for4differeny4conducyor48nd4PM4yhicc neffef.4

	μ_CO0OmmO	μ_CO2OmmO	μ_CO5OmmO	μ_CO%OmmO	μ_CO10OmmO
μ_{FC}O5OmmO	44NONs/mO '0.%ONg)O	215/ NN0ONs/mO '0.8%ONg)O	244/ 2%%ONs/mO '1.20Ng)O	229/ 2440Ns/mO '1.40Ng)O	205/ 22NONs/mO '1.%%ONg)O
μ_{FC}O%OmmO	%1NONs/mO '10Ng)O	N%/ 4040Ns/mO '1.20Ng)O	409/ N010Ns/mO '1.5NONg)O	N%8/ 2%%ONs/mO '1.%%ONg)O	NN4/ 2N%ONs/mO '2.120Ng)O
μ_{FC}O10Omm	9N5ONs/mO '1.4%ONg)O	%%0/ 5200Ns/mO '1.%ONg)O	%%N N500Ns/mO '2.040Ng)O	%11/ N000Ns/mO '2.NONg)O	5N2/ 25%ONs/mO '2.%ONg)O
μ_{FC}O12Omm	O19NONs/mO '1.80Ng)O	890/ %00ONs/mO '2.0NONg)O	852/ N%9ONs/mO '2.40Ng)O	%%5/ N180Ns/mO '2.%%ONg)O	%%0/ 2%%ONs/mO 'N0%ONg)O

O

IHDo0zS0Op0o%pDHOoppXnumODoHzuDpo0ODHzOoupS0OmDgHSpsOphXDNHSssSs1OphSOgSHS0DpSzOzDmpXHgO DoS00XDSHp0o0OzX0S0SHpOmDgHSpOphXDNHSssSsOXOp8oppSzOEs.OphSODoHzuDpo0OphXDNHSssOHOFXg.O5-10.OpOXO 0SESDSSzOphDpOphSOmDXnumO2DmpXHgODoS00XDSHpOXO%pDHSzODpODoHzuDpo0OphXDNHSss$0\mu_C$OSHuD8OpoO2Omm1O 0SgDDz8SssOo0OhSOoupS0OmDgHSps'OphXDNHSss$0\mu_{FC}$.O0O

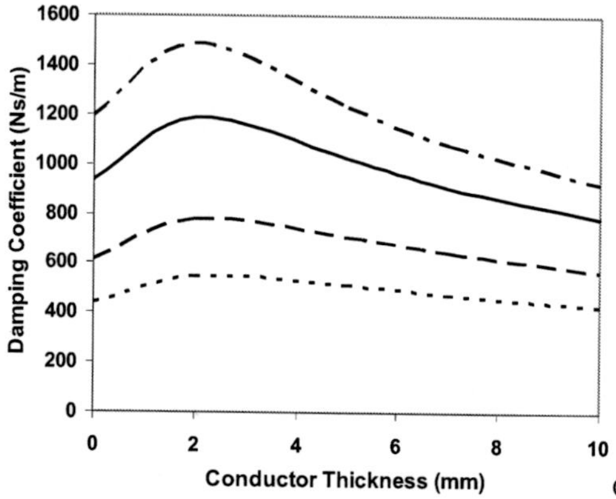

O

: iA4/-904D8A pinA4l oefficieny4vf.48luA inuA 4conducyor4yhicc neff4for4differeny4ouyer4A 8Aneyf4 yhicc neffef.4□ □□ :4σ_{OB} ⌋ $12GG$,4□ □ :4σ_{OB} ⌋ $10GG$,4□ 4□ :4σ_{OB} ⌋ $\%GG$,4□□4σ_{OB} ⌋ $5GG$.4

– sOphSODzzSzO SXghpOzuSOpOphSOSzzyODu0OSHpODssXESOu%syspSmOXOo0OhXghOXnpo0pDHDSOHOphSO2SsXgHO p0oDSzu0S1OphSODpXOo0OphSO2DmpXHgODoS00XDSHpOEs.OphSODzzSzO SXghpOpOphSOsyspSmOXOD8soOp8oppSzOHO FXg.O5-11ODoHDXmXHgOXnXDOOoppXnumOphXDNHSssOo0OphSODoHzuDpo0.O0O

– sOX8uspODpSzOXHOFXg.O5-111OphSOzDmpXHgODoS00XDSHpOzSHsXyOs8XghpOyOXHODSDsSsO XphOphSOoupS0O mDgHSps'OphXDNHSss.O 8houghOphSOoupS0OmDgHSps'OphXDNHSssOhou8zOSOSSDpSzOXDSSzOHOphSOSHuXSzO zDmpXHgODoS00XDSHpOXpsOoppXnumODDuSO0Ssu8pXHgOXHODOXghSO2DmpXHgODoS00XDSHpO2SHsXpy)OXO%pDHSzO ~15OmmO0o0ODuumXHum)ODsOho/ HOXOFXg.O5-12.O HopOphSOXnpo0pDHpODoHDusXoHO0omOphXOXguOSOXOphDpO phSODoHzuDpo0OmDpSOXDSOOSDpOXOHOuSHDSzOyOphSOgSomSpOyOo0OoupS0OmDgHSps.OpOXO0SESDSSzOphDpOphSO oppXnumOmDgHSpOphXDNHSssO2SpSHzsOHOphSOSSDpSzODoHzuDpo0OmDpSOXDOFo0OHspDHDSODompDXHgOphSO p/ oOmDpSOXDsOo0ODuumXHumODHzODoppSO%OphO XphOXghODoHzuDpXEXpy/zSHsXpyODpXo1OXOXOho/ HOphDpO XphO DOmDgHSpOphXDNHSss$0\mu_{FC}$OXghSOphDHO10mm1ODoppSOXOmo0SODppSDXHgO2uSOpOXpsOXghSOgSHSODpXoHOo0O zDmpXHgOo0OSO2SHsXpy.O

O O 1N%O

O

j DsSzOHphSOSHuXSzpDsXESO2DmpXHgODoSOXXSHpOo0phSOhy%OXzO2DmpSOSsXgHO~1400ONs/m)1OphSO p/ oOpopSHpXOOoHzuDpo0OhoXSsOOSOnSHpXoHSzOHOOD%SO-5.OoHSHuSHpy1OphSODoppSOODoHzuDpo0O XphO oupSOOnDgHSpOphXONHSssOα_{FC}OSHuDOpoO10OmmOXOSSSDpSz1OXnp0oEXHgOphSO2DmpXHgODoSOXXSHpOOHzOphSO zDmpXHgODoSOXXSHp/ SXghpO2SHXpyO%OOHzO2%1OSspSDpXESSy.O

T8ble47-744Tf o4opyion4or4yhe4conducyor4A 8yeri8l48nd4ize.4

8 DpSOXO8O	OupSOO8 DgHSps'OOhXONHSss	DDmpXHgODoSOXXSHpO	WSXghp
– 8umXHumO	12O	1490O	2.0NO
OoppSOO	10O	15%0O	2.08O

O

OWXphOphSOXHD8O2SsXgHSzODoHzuDpo0OOHzOoupSOOnDgHSps1OO15%0ONs/mOpDsXESO2DmpXHgODoSOXXSHpOXO p0oEXSzOpOphSODospOo0OzXHgO2.1ONgOo0OzXpXoHD8O SXghpO- sOOOSsu8p1OphSOXHD8OzXnSHsXoHsOo0OphSO hy%OXzOS8SDp0omDgHSpXOhyzODu8XO2DmpSOXO%pDHSzOHzOXpSzOHOOD%SO-%O

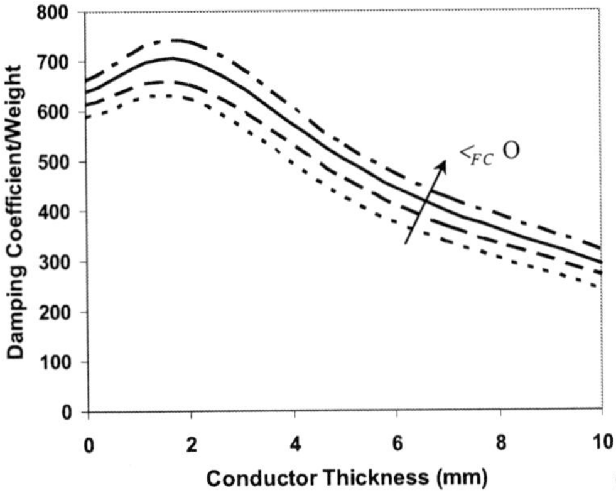

O

: iA47-9944D8A pinA4coefficieny/f eiAhy4vf.48luA inuA 4conducyor4yhicc neff4for4differeny4ouyer4 A 8Aney4 yhiccneffef.4 □ □□ :4 σ_{OB} ⌋ 12GG,4 □ □ :4 σ_{OB} ⌋ 10GG,4 □ 4 □ :4 σ_{OB} ⌋ %GG,4 □□□4 σ_{OB} ⌋ 5GG.4

O

O

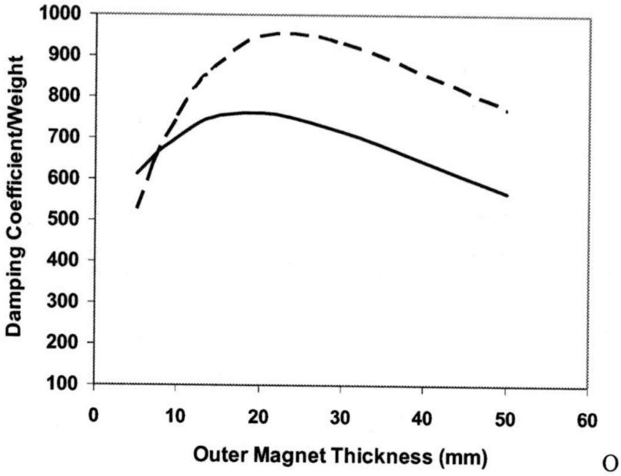

O

: iA47-9244D8A pinA4coefficieny/f eiAhy4vf.4ouyer4A 8Aney4yhicc neff4Pξ_c µ $2GG$)4for4□ 4□ :4 d opper48nd4□ □ :48luA inuA .4

T8ble47-644: in814liA enfionf4of4yhe4hybrid4elecyroA 8Aneyic/hybrid4d8A per.4

OupS0OnDgHSpsOD0DyO	
Num%S0Oo0po8SsypO	4O
Po8SOpXDhpO	24OmmO
8 DgHSpOhXDNHSssO$_{z}$ O	12OmmO
8 DgHSpOupsXzSO2XDmSpS0O	114Ommo
OhXDNHSssO$_{FC}$O	10OmmO
OopDX8SHgphO$_{e}$O	9%OmmO
OupS0OnDgHSps'O SXghpO	1.520NgO
8 DgHSps'OnDpS0XD8O	Nz-FS-j 1O $_{2}$O1.1%O0O
OoHzuDpo0$_{F}$	
OoHzuDpo0'sOnDpS0XD8O	OoppS0O0
OoHzuDpXEXpyO0	5.9S%OS/mO
DSHsXpyO	8.9SNONg/m^{N}
OhXDNHSssO$_{C}$O	2OmmO
OoHzuDpo0O SXghpO	0.5%ONgO

O

O O

O

IpOhou8zO'SOnSHpXoHSzQhDpOOzzXHgOoupSOOP8 sOHDOSDsSsQhSOnDgHSpXOO8uxO2SHsXyyQDsXHgQhOoughO phSOpDpoOODoX8s1OSHhDHDXHgQhSOpSOobOmDHDSOoOQhSODpXESOu%syspSm.OOompDOXHgQhSOhy%OXOzDmpSOO / XphODHzO XphoupOoupSOOnDgHSpsOho/ sOOO.004OOO2%)OHDOSDsSOHOnDxXnumOO8uxO2SHsXyyOHQhSOpDpoOO DoX8s1O hXOhOXOObpOXgHOXOXODHp1ODHzOODHO'SOOHSg8SDpSz.O

OhSOOHDO2SsXgHOpSDXOXODpXoHsOoOQhSO8SDpObmDgHSpXO'SzzyOhy%OXOzDmpSOODOSOXpSzOHOODO'8SO5-%O OhSOo%pDHSzOpSDXOXODpXoHsO X8O'8SOOmpDOSzQoQhDpOOQhSOphSOO8SDpObmDgHSpXOhyzODu8XOO2DmpSOOHO phSOHSxpOSDpXoHO

T8ble47-744P ver8ll4Specific8yionf4bf4he4propofed4elecyroA8Aneyic/eddy4hybrid4l8A per.4

p uDHpXyO	– DuSO
PDsXESO2DmpXHgODoSOOXOXSHpO	15%OONs/mO
ExpSOHDO2XOmSpSOOO	115OnmO
OopDXO8SHgpho	2%OOnmO
OopDXO SXghpO	N4/ 2.1O5.5ONgO

O

7.6 Ein8l4CoA p8rifon4bf4HybriC4D8A perf48nC4Conclufionf4

– OpSOQhSODpXESOE8SDpObmDgHSpXOODmpSOOED)OXQpObposSzODHzO2SsXgHSzOHOOhDppSO41QhSOHS/ OXSDOoOO DzzXHgQDsXESO2DmpXHgQoQhSOEDOXQpOSsSHpSzODHzO2SES8pSzOHQhXOOhDppSO.OpOXOSxp8DXHSzOho/ QhSO hyzODu8XO'SzzyODuOOSHpOoODSsODDHpOoEXSOOOnXHXnumOSHuXOSzQDsXESO2DmpXHgQoOSHsuOSOOmD8SO1O 8XghpSO1ODHzOnoOSOS8XO'8SODpXESO2DmpSOO XphOSSssQpo/ SOODoHsumppXoHOOhSOSDsX'X8XyOoOQhSOhy%OXO zDmpSOODoHDSppOXOpuzX8z1OpDpXHgO XphOS8SDpXHgOOppObpOXOpSOhy%OXOzDmpSOQopo8gX8s1ODo8b/ SzO'yO XHpXOXOX2XHgOoOQhSO8SDpObmDgHSpXOOu%syspSm.OO/ oO2XOSOSHpQopo8gX8sODSQpObposSzODHzO2SsXgHSzO %DsSzOoHOphSOzSOESzOzSsXgHOOSHuXOmSHpsOXHOOhDppSOO2.OOhSOOXspOpObposSzOhy%OXOzDmpSOO DoHDXguODpXoHupXOXSsOhyzODu8XOO2DmpXHgOSOOSDp1O hXOSphSOSDoHzOoHSOusSsQhSOzzyODuOOSHpO2DmpXHgO phSHomSHoHO

Q opOhy%OXO2DmpSOO2SsXgHsOhDSzOXnX8DOODpXESOompoHSHp1O hXOhOXO2SsXgHSzOHOSSDpXoHO5.2.1 1O pObEXXHgQhSOnXHXnumODpXESOobODSOSHuXOSzOboOOhy%OXOzDmpSOs.OpOXOho/ HQhDpODsQhSOOSHuXOSzO DpXESOboODSOXOSzuDSzOOomOlN0OONOHOOu8yODpXESOODsSOpoON5OONOHOphSOhy%OXOysypSm1QhSOpo/ SOO DoHsumppXoHOXOSzuDSzO4%OHQhSOhy%OXO2DmpSOODompDOSzQoQhDpOOODpXESOED.OOhXODoOSspoHzsQpoO

O

DO%1%OSzuDpXHOXHphSO SXghpOoOphSODpXESOuHpOhSOXHDO2SsXgHpSDXOXDpXHsOoOphosSO oOhy%OXO zDmpSOsOOSpOSsSHpSzOHzOompDOSzpoOphDpOOphSOEDO2DmpSOXHOOD%SO-8.O

FoOphSOXOspOhy%OXO2DmpSO2SsXgHOXHO hXOhphSOhyzODu8XOOoODSsOOSOupXXSzpoOpOoEXSphSOSHuXSzO pDsXESO2DmpXHg1OOgSHSOD8OhyzODu8XO2DmpSO2SsXgHOnSphozOXpOSsSHpSzOHzOphSHOHDOo/ SzO2o/ HppoO phSOpSDXOXODDsSOoOphSOSSDpOomDgHSpXOhyzODu8XO2DmpSO2SsXgHOpOXO2SmoHspODpSzOphDpOOXgHXOXODHpO pDsXESO2DmpXHgOoOO-1%O0ONs/mODHO%SOpOoEXSz1O%yOphSODzzXpXHOoOphSOEXDousO8uXOpoOphSODpXESO zDmpSO1O hXOhOguDOHpSSsODODX-sDOSO2DmpSOXHODDsSOoOOSSDpOoHXOODX8uOS.OHODzzXpXHOphSOhyzODu8XO syspSmOhDsOphSO%SHSOXpOoODzzXHgOODOo8XHgOysypSmOpoOphSO2DmpSO.O- moHgO2XOSOSHpO2DmpSOsO8XpSzOHO OD%SO5-8 1OXpOhDsOphSO8o/ SspO SXghpOOhSOzDmpSOODSoOSxhX%XpSOODsXnX8DOOzDmpXHgO%ShDEXOOXHO SxpSHsXH/DompOSssXHOpDgSs1O hXSOphSO2DmpXHgOoSOOXOXSHpOXHODompOSssXHOXO8XghpyO8o/ SO2uSOpoO phSOSsSOEoXOgDsODompOSssX%XXpy.OOompDOXHgOphXO2DmpSO XphOphSODpXESOED1OXOXOSESDSSzOphDpOphSO / SXghpODHzOpo/ SOODoHsumppXHOoOOphSOzSsXgHSzOhy%OXOzDmpSOOXOSzuDSzO%yO5%8%ODHzO%4%1O OSspSDpXES8y.O

T8ble4 7-84: in8l4 coA p8rifon4 bf4 yhe4 f o4 hybrid4 l8A per4 confiAur8yion f4 8nd4 yhe4 8cyive4 D.4

SpSDXOXDpXHO	Hy%OXOHyzODu8XO	Hy%OXOEzzyODu0SHpO	– DpXESOEDO	DSsXgHODXpSOXO
OopD8O SXghpO	5.2ONgO	5.5ONgO	11.8ONgO	<%ONgO
OopD8SSHgphO	N00OmmO	2%O0OmmO	N50OmmO	<%O0OmmO
ExpSOHDO2XOmSpSOO	94OmmO	115OmmO	120OmmO	<150OmmO
PDsXESO2DmpXHgOoSO.O	1%O00ONs/mO	15%O0ONs/mO	550ONs/mO	~1400ONs/mO
PSDNODpXESOoODSO	N20ONO	N20ONO	1200ONO	1N00O&ON50ONO
OSHuXSzpo/ SOC4OuHpO	N80OWO	N80OWO	1480OWO	O

FoOphSOSDoHzOSSSDpOomDgHSpXOSzzyOhy%OXO2DmpSO1OphSOzXnSHsXHsOoOphSODzzSzODoHzuDpoOOHzO mDgHSpsOOSOSSDpSzOpoOmXHXnXSOphSO SXghpODzzSzOpoOphSOsyspSmO hXSOmDXpDXHXHgOphSOSHuXSzO pDsXESO2DmpXHg.OpOXOho/ HpphDpOO15%O0ONs/mOpDsXESO2DmpXHgOXOOhXESSzO XphOphSOHpOozuDpXHOoOO phSOSzzyODu0SHpO2DmpXHgOSOOSDpOpoOphSOhy%OXOysypSmOpOphSODospOoODzzXHgO2.1ONgOoODzzXpXHD8O SXghpO OompDOXHgOphXO2SsXgHO XphOphSOoOmSOpOoposSzOhy%OXO2DmpSO1OpsO SXghpOXOHDOSDsSzO%yO5%ODHzOpsO oESOD8OSSHgphOXOSzuDSzON%1O hXSOphSyOhDESOOXnX8DO2DmpXHgOpSOoOmDHDS.OO

– HOoESOD8OompDOXoHOoOphSODpXESOSSDpOomDgHSpXO2DmpSO XphOphSOpOoposSzOhy%OXO2DmpSOsOHO phXOSDpXHOOSESDS8OphDpOphSO8DpSOOhDsO%4%OSssOpo/ SOODoHsumppXHODHzO-55%OSssO SXghp1O hXSO mDXpDXHXHgOphSODpXESO2DmpXHgOpSOoOmDHDSODzEDHpDgSs.OO

ChLpyer 4 64

SuA A Lry 4 ow Rewulyw Ln C 4 Guyure 4 Worc 4

[illegible]

[illegible]

[illegible]

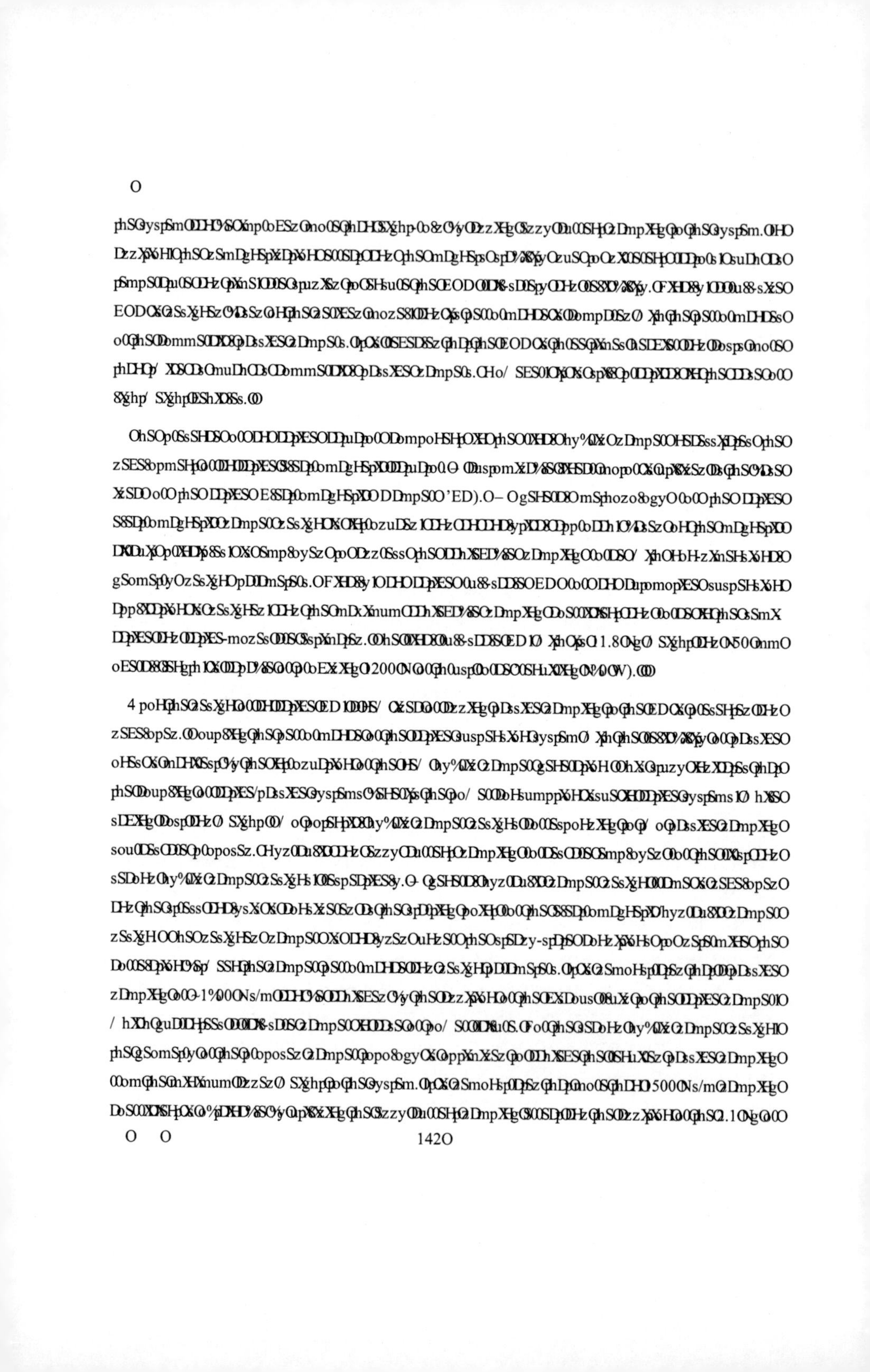

O

/ SXghpOpoQhSOysp$m.OOompDXHgQhSOu8y-DDpXESOEDO XphQhSOHS/ 8yQpOoposSzOhy%XOaDmpSOaSsXgHsO XnpSSsQhDpO SXghpODHzQpo/ SOODoHsumppXHoOQhSOEDODou&zO%SOSzuDSzOmoOSQhDHO0%ODHzOO%1O OSspSDpXESSy1O%yODoHESOpXHgOXOHpoODhy%OXaDmpSO.OO

6.9 Euyure4f orc4

j DsSzOoHOphSODomp8SpSzO/ oONOoHODDpXESOSSSDpOomDgHSpXOOzDmpSOs1OphSOOo8o/ XHgOOSsSDOOhOXO OSDommSHzSzOoOOupuOSOpuzy.O

- OHpOODpXOSOXO DsOSSHQhDpQhSOusSOoOOSsSOEoXOgDsOOhDm%SOOHQhSOOXspOhy%OXaDmpSOaSsXgHODDusSsO phSOOsymmSpOXO%ShDEXoOOHODomp0SssXoH/SxpSHsXoHO hXOhO DsOHbpOaXDussSzOHaSpphOHQhXO%oNO IpOXOuggSspSzQhDpOOO8uXzOHDsysXOXQpSOOoOmSzQpoOmozS8QhXOSOOSDpOoQhSOhDOODpSOXpXDsOoQhSO OSsSOEoXOOhDm%SOO XO%SOaSsXgHSzOHODOuDhO DyQhDpQhSOaDmpSOOpopSHpXO8yOSxhX%pQhSOaSsXOSzO DsymmSpOXOaDmpXHgO%ShDEXoOOHODomp0SssXoH/SxpSHsXoHOpDgSs.O
- OOhSOpDsXESOzDmpXHgOpSOoOmDHDSOoOOphSOSSSDpOomDgHSpXO/hy%OXOzDmpSOOzSsXgHOXOhXgh8yO XHO8uSHDSzO%yQhSOO8uXOEXDosXpyODHzOsoO XphQhSOopSODpXHgOpSmpSODpuOS.O-pQhSOsDmSQpXnS1QhSO EXDousO8uXzODpsODsOOOo8XHgOysp$mOoOQhSOpDpoOODoXs.O sOOpDpOoOOupuOSO oONIQhSOSOOSDpOoOQhSO moEXHgO8uXzOHQhSOpSDzy-spDpSO oONXHgOpSmpSODpuOSOoOQhSOysp$mODou&zO%SODHDyzSz.O-8houghO phSOopSODpXHgOpSmpSODpuOSOXOXHO8uSHDSzO%yQhSOopSODpXHgODHzOOoDzODoHzXpXoHsODsO SO8ODsQhSO SHEXoHmSHpDOpSmpSODpuOS1QhSOhyzODu8XOOhy%OXOzDmpSOOzSsXgHODou&zO%SOmozXOXSzO%DsSzOoHODO moOSOODDuODpSOopSODpXHgOpSmpSODpuOS.OO
- OOhSOzSsXgHODHzOmozS8XHgOoOOophSOOhy%OXOzDmpSOODoHOXguODpXoHsOXOODHbphSOONSyOOupuOSO/ oONO suggSspXoHOO%oOSxDmpS1QhSOOXspOhyzODu8XOOhy%OXaDmpSOaSsXgHODou&zO%SOpopSHpXO8yOmozXOXSzQpoO DOp/ XHpu%SOSSSDpOomDgHSpXO/hy%OXOzDmpSO1O hSOSQhSOOSsSOEoXOXOODoHDSHpOXOD8yQp8DDSzOOOouHzO phSOmDXHODy8XHzS.OOhXODDusSsQhSOzDmpSOQpoOopSODpSO XphO%o/ SOOHXpXO8OSsSOEoXOgDsQpOSssuOS1O / hXSSQhSOaDmpSOOXOSsssOusDSppX%SQpoOaDmDgSODDusSzO%yOXnpDDpOO

- O-HbphSOOSDommSHzDpXoHOXQpoOzSES8opODHOXHzuspOXO8OESOsXoHOoOQhSOzSES8opSzOhy%OXOzDmpSOO zSsXgHs.OOhSOSSSDpOomDgHSpXO/hyzODu8XOOhy%OXOzDmpSOODoHDSppODou&zO%SOXnp8SmSHpSzODsODO mozu8DOODompoHSHpQpoO%SODzzSzQpoQhSOO-phS-shS8OQpDsXESOaDmpSOs.O

AppenCix4A4

Review4bw4Glecyro A LAneyiwA 4

A.9 MLxwell9w4GquLyionw4

E8SDp0omDgHSpX00X8&zsCD0SQgoES0HSzO⅔yC8 Dx/ S8'sCSHuDpX6Hs1O' hS0S0*H*O'mDgHSpX000X8&zOXHpSHsXpyO '– /m))1O*H*OmDgHSpX00xO2SHsXpyOO))1O*H*OE8SDp0X00X8&zOXHpSHsXpyO– /m))1ODHzO*H*OE8SDp0X00xO2SHsXpyO 'O/mN))OD0SQhSO0X8&zs.O*D*' *0* YG^2) ODHzOσ' ν YG^N) OD0SQhSOou0DSQpS0ms.O

O $$+\times \boldsymbol{D} \rfloor \boldsymbol{D} + \frac{\rfloor \boldsymbol{D}}{\rfloor t} 1O$$ '– -1)O

O $$- = \boldsymbol{D} = 0\ 1O$$ '– -2)O

O $$+\times \boldsymbol{D} \rfloor + \frac{\rfloor \boldsymbol{D}}{\rfloor t} 1O$$ '– -N)O

O $$)\ \cdot \boldsymbol{D} \cdot\ \cdot .O$$ '– -4)O

OhX0ysp8mOo0CSHuDpX6Hs1O' hXOhOD0SOHpOD8OHzSpSHzSHp1O0Sp0SsSHpsOD0ysp8mOo0O1 2OuHNHb/ HO0X8&zO DompoHSHps.OFo0ODDompS8pSQhSo0yOo00X8&zOSHuDpX6Hs1QhSODzzXpX6HDX0DoHspXupXESOSHuDpX6HsOnusp0⅔SO DzzSzQpo08 Dx/ S8'sOSHuDpX6HsOD8O0o88o/ s:O

O $$\boldsymbol{D} \rfloor \sigma_0'\boldsymbol{D} + \boldsymbol{n}\ )\ 1O$$ '– -5)O

O $$\boldsymbol{D} \rfloor \sigma_0 \boldsymbol{D} + \boldsymbol{n}\ 1O$$ '– -%)O

DHzO

O $$\boldsymbol{D}\, \varepsilon\ \varepsilon\ \boldsymbol{D}\ 1O$$ '– -%)O

/ hS0S0*P* O'– /m)OX0QhSOmDgHSpXDpX6H1O*P*O'O/m^2)OX0QhSOpo8D0XDpX6H1O $\sigma_0 \rfloor$ U$\sigma \times 10^{\%}$ 'Om/–)1O $\sigma_0 \rfloor 8.85U\times 10^{)12}$ 'F/m)1ODHzO σ '– /– m)OD0SOQhSOpS0mSD⅔X8Xpy1OpS0mXpXEXpy1ODHzODoHzuDpXEXpy1O 0SspSDpXES8y.OD

8 Dx/ S8'sOSHuDpX6HsODDH9⅔SO0xp0SssSzOXHQhSOXHpSg0D8O0o0mO⅔yQDNXHgQhSOu00DDSOXHpSg0D8O0QhSODu08O EHs.O– -1)ODHzO– -N)1QhSHODpp8yXHgOSpoNSs'QhSo0SmQpoQhSOSS0p-hDHzOX&S.O- 8so1QhSO2XES0gSHDSOEHs.O '– -2)ODHzO– -4)OD0SOXHpSg0DpSz1ODHzQhSHQhSO2XES0gSHDSQhSo0SmOX0Dpp8X8zQpoQhSOS0p-hDHzOX&S.O

O

O $$\oint_v D = \!/\!l = \int_v (D - \frac{=D}{=t}) = \!/\!D1O$$ '– -8)O

O $$\oint_v D = \!/\!l = -\int_v (\frac{=D}{=t}) = \!/\!D1O$$ '– -9)O

O $$\oint_v D \rfloor 0D \rfloor\ 0\ 1O$$ '– -10)O

O $$\oint_v D \rfloor 0D \rfloor\ \varepsilon \sigma 0 \rfloor .O$$ '– -11)O

'– -8)OXhSQgSHSODOo0mOo0O- mpSOS'sOD' 1O– -9)OXOFDODzDy's OD' 1O hXOhOspDfsSQhDpQhSOH-0X&zOXO gSHSODpSzO%yODpXnSOEDOyXHgOnDgHSpXDO0XS&z1ODHzO– -10)OspDfsSOhDpQhSOoup' DOzO0uxOoESOODODoSSzO su0ODSOX2SOo1OhK1OhSOSODSOHbOnDgHSpXDOnoHopo8SsOXo8DpSzOXHNsOo0Oou0DSs).OOhSOSDspOSHuDpXoHOXO DD8SzOo Duss'sOD' 1ODHzOXpSpDfsSOhDpQhSOopDOoup' DOzOSSSDpOXDO0uxOoESOODODoSSzOu0ODSOSHuD8sOhSO 0SSSOhDgSsOSHDoSSzO%yOhSOu0ODS.O

O

A.2 I oyenyiLlw4

– OESDpo0OpopSHpXD8ODHzOOLDDODOpopSHpXD8OODSOESDpo0OX&zs1O hosSODu08ODHzOHSgDpXESOg0DzXSHpODSODO gXESHOESDpo0OX&z1OSspSDpXES8y.OOypXDD8y1OXO Gno0SODoHESHXSHpOo%pDXHOhSOX&zsO%yOusXHgOpopSHpXD8O 0uHDpXoHs.O- SDpo0OH)ODHzOsDDODOpopSHpXD8sO σ)ODSOusSzOpoODoHESOpO8 Dx/ S8'sOou0ODoupSSz1OXsp- o0zSOSHuDpXoHsOHpoOp/ oOuHDoupSSzOsSDoHz-o0zSOSHuDpXoHs.OOhSOESDpo0OpopSHpXD8OHOXOzSOXHSzO%yO '– -2)ODsOo88o/ s:O

O $$+ \rfloor A \rfloor 0 \Rightarrow A \rfloor + \times A .O$$ '– -12)O

OhSOsDDODOpopSHpXD8O σ OXO2SOXHSzO%yO– -N)ODsO

O $$+ \times \rfloor A + \frac{\rfloor A}{\rfloor t} + \rfloor 0 \Rightarrow A + \frac{\rfloor A}{\rfloor t} \rfloor ++ \sigma .O$$ '– -1N)O

j yOu%spXupXHgO– -12)ODHzO– -1N)OHpoOhSOSmDXHXHgOp/ oO8 Dx/ S8OSHuDpXoHs1O

O O

O

$$+^2A++'+\rfloor A+\sigma\sigma\frac{\rfloor\sigma}{\rfloor t})+\sigma\sigma\frac{\rfloor^2A}{\rfloor t^2}\rfloor+\sigma A1$$
$$+^2\sigma+\frac{\rfloor'+\rfloor A)}{\rfloor t}\rfloor+\frac{\sigma}{\sigma}.$$
O '– -14)O

j yOxnposXHgOphSO o0SHzQgDugSODoHzXpXoHO+ $\rfloor 2\ \rfloor+\sigma\sigma\frac{\rfloor\sigma}{\rfloor t}$)OphSOp' o0OXHDXoSDoHz-o0zSO0SHuDpXoHsO0OSO

$$+^2 2+\sigma\sigma\frac{\rfloor^2 2}{\rfloor t^2}\rfloor+\sigma 2\,1$$
$$+^2\sigma+\sigma\sigma\frac{\rfloor^2\sigma}{\rfloor t^2}\rfloor+\frac{\sigma}{\sigma}.$$
O '– -15)O

8 Dx/ SS8'sOSHuDpXoHsOspDHOphSOSHpXOSOSSDp0omDgHSpXOODpp8XODpXoHs1O hXOhOOOSOODpSgo0XzSzOOsOhXghO 00SHuSHDy1OHuDsXspDpXO1OODHzOspDpXODOOhSOHuDsXspDpXOOphSo0yOgoESOHsOOO' XzSOODHgSOo0ODpp8XODpXoHs1O XHDuzXHgOSSSDp0omSDhDHXODXaSEXOSsOOHzOSzzyOOu0OSHpOphSHomSHDO

IHOphSOHuDsXspDpXOODpp0oxXnDpXoHIOphSOzXsp8ODSmSHpOOu0OSHpOpS0mO=D Y=t OXOXgHo0SzOXHO– -1).OOhSO popSHpXOXSHuDpXoHsOOOSOOSzuDSzOpoOPoXsoH'sOSHuDpXoHs1O

$$)\ ^2 2\ ]\ )\ J\,2\,1$$
$$)\ ^2 J\ ]\ )\ \frac{J}{J}.$$
O '– -1%)O

IHO 00SSO spDDS1O phSO so8upXoHO poO '– -1%)O XO SxpOSssSzO %yO usXHgO o0SSH'sO 0uHDpXoHO $m'2\,12])\ ]\ )\ '1\ YUJ\ )'1\ Y|2\ )\ 2|)$ OboOOopSODpoOO+ ^{2} :O

$$2'\,21t)\ ]\ \frac{J_0}{UJ}\int_{]}\frac{2'\,2]1t)}{|2\ )\ 2|}V]\ ]$$
O '– -1%)O

DHzO

$$J'\,21t)\ ]\ \frac{1}{UJJ_0}\int_{]}\frac{J'\,2]1t)}{|2\ )\ 2|}V]\ ]1$$
O '– -18)O

/ hSOSOaOXOphSO%sSOEDpXoHpoXHpOODHzO2]OXOphSOou0DSOpoXHpO

O

OhSOspDpXD-0XS8zQhSo0yOXODSoODpp8XSzQpOODOEDOXSpyOb0ODpp8XODpXoHsOsuDhODsQpS0mDHSHpOmDgHSpO spOuDpuOSs1ODHzOXSDpODu0OSHpO2SEXOSs.OHQhSOspDpXD-0XS8zQhSo0y1O%phQXnS-zSpSHzSHpOS0msODDSOXgHb0SzO 'εD Yεt ε εD Yεt ε 0).OOhSOso8upXoHsOb0QhSOpopSHpXO8OSHuDpXoHsODOSOXnXDO0QpoQhosSOHOhSOHuDsX spDpXOODsS1OuHSssQhSOsDDXD0ODHzOESDpo0QpopSHpXDsOD0SObpQXnSO2SpSHzSHpOO

O

A.9 MLAneyic4GielC4AnLlywiw4

OHDSOHOXO2SpS0mXHSzO%yODDXDu8DpXHgO--1%)1QhSOmDgHSpXOO8uxO2SHsXyOHODDHO%SODompupSzO%yOusXHgO 2]) × 2 1ODHzO

$$2\,'\,2\,1t)\,]\;\frac{J}{UJ}\,]\,\frac{2\,'\,2]1t)\times'\,2\,)\;\;2])}{|2\,)\;\;2|^{N}}W]\,].\mathrm{O} \qquad '--19)\mathrm{O}$$

Fo0ODQhXHOXSDmSHpO/ XS)1$2V$]]) EV] 1ODHzO--19)ODoHES0pSzQpoQhSO XopSDEDQpSD' O

$$2\,'\,2\,1t)\,]\;\frac{JB}{UJ}\oint_{B}\frac{W]\times'\,2\,)\;\;2])}{|2\,)\;\;2|^{N}}.\mathrm{O} \qquad '--20)\mathrm{O}$$

Fo0Osu0ODDSODu0OSHps1O2W]]) $2W$] 1O--19)O%SDomSsO

$$2\,'\,2\,1t)\,]\;\frac{J}{UJ}\,]\,\frac{2\,2]1t)\times'\,2\,)\;\;2])}{|2\,)\;\;2|^{N}}W]\,1\mathrm{O} \qquad '--21)\mathrm{O}$$

/ hSOSOHO--/m)OXQhSOsu0ODDSODu0OSHpO2SHsXy.O

OhSOSOD0SODOHum%SOOb0ODpp0bDDhSsObOODDXDu8DpXHgQhSOmDgHSpXOOXS8zOb0ODQpS0mDHSHpOmDgHSpOOhSO sXnp8SspODpp0bDDhOXQhSOzXpo8SOnomSHpOnozS8QhDpOXODpp0bpOXDpSOb0QhSOOXS8zODDXDu8DpXoHODpOoHgO zXpDHDSsO0bomQhSOmDgHSp1ODompDOSzO XhQhSOmDgHSpOzXOmSpS0.OOhSODhDOgSOnSphozODHzQhSODu0OSHpO mSphozOnozS8QhSOmDgHSpO XhODO2XpOX%upXoHOb0QhSODhDOgSODHzODu0OSHp1OSspSDpXES8y.O

O

A.4 MLAneyic4Dipole4MoA eny4MeyhoC4

IHQhXOnSphoz 1QhSQpS0mDHSHpOmDgHSpOXOnozS8SzO%yODDXHg8SODu0OSHpODD0yXHgO8op1ODsOho/ HOXOFXg.O --O).O

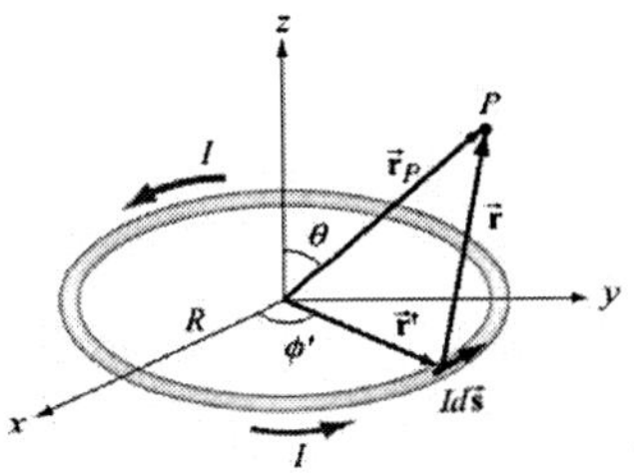

Fig. A- 1 A sing)Dc1rrDn(carrying)oop.

"k" zpplking" the" "ict-Szvzrt" lz9" (A-,")" bk" $dz'\,'\;Rd\phi'\hat{\phi},\ \boldsymbol{x}-\boldsymbol{x}'\,'\;,\hat{z}-R\hat{r}$"

zn" $|\boldsymbol{x}-\boldsymbol{x}'|\,'\;\sqrt{R^{\cdot}+,^{\cdot}}$,'the'I zgnetic'flux'zlcng'the'zxiA'cf'the'current'lccp'reAultAin"

$$J(,)\,'\;\frac{\mu_{\cdot},R^{\cdot}}{4\mu}\,\mu^{,\pi}\frac{d\mu'}{(R^{\cdot}+,^{\cdot})^{5/,}}\hat{z}, \quad \text{(A-,,)}$$

9hich'iA're"uce""tc""

$$m(,)\,'\;\frac{\phi_{\cdot}\boldsymbol{m}}{,\phi(R^{\cdot}+,^{\cdot})^{5/,}}, \quad \text{(A-,5)}$$

9here"$\boldsymbol{m}$ μ $I\mu R^{\cdot}\hat{z}$'iA'the'I zgnetic""ipcle'I cI ent'cf'the'lccp/'AA'it'9ill'be"Aeen'in'the'next"Aecticn," the'equivzlent""ipcle'I cI ent'fcr'z'I zgnet'iA'$\boldsymbol{m}\,'\;,\;l\phi R^{\cdot}\hat{z}$,'9hen'z'ccnAtznt'I zgnetizzticn'in'the'M "irecticn'iA'zAAuI e"/'R'zn""l'repreAent'the'rz"iuA'zn""length'cf'the'I zgnet/"

A.5 ThDC1rrDn(MMD) (Ma5nD(ic F)1x Ca)c1)a(iMn)

In'the"current'I c"el,'the'I zgnet'iA'expreAAe""zA'the""iAtributicn"cf'the"equivzlent"current,'9hich'iA' uAe""zA'z"Acurce"terI "in"the"I zgnetcAztic"fiel""equzticnA""k"AubAtituting"(A-",)"intc"(A-"),"zn"" iI pcAing'the'Cculcl b'Gzuge'('/J'^),"

$$'\,{}^{\cdot}\boldsymbol{m}'\;-\phi_{\cdot}(\boldsymbol{m}+'\;\times\boldsymbol{x}\;)/ \quad \text{(A-,4)}$$

”

(A-, 4)” czuAeA’ the” “efiniticn” cf” the” I zgnetic” vcluI e” current” “enAtk” m_M ' ' ×x /” If” the” I zgnetizzticn’M’iAreAricte“’tc’z’vcluI e” ,’zn“’fzllAtc’zerc’cutA“e’cf” ,’(A-”9)”iAre“uce“’tc”

$$m(x)' \frac{\phi}{4\phi}\Gamma\frac{m_M(x')\times(x-x')}{|x-x'|^5}d' ' + \frac{\phi}{4\phi}\Gamma\frac{r_M(x')\times(x-x')}{|x-x'|^5}ds' , \quad (A-, 5)$$

9 here”m_M’zn“” r_M’zre’equivzlent’vcluI e’zn“’Aurfzce’current”“enAtieA’zn“”“efine“’bk:”

$$\begin{aligned} J_M &' ' \times x \;(, /, ') V, lR, eRR, eRVdeRs, Vy, \\ j_M &' x \times \hat{n} \;(, /,) VR, fVRe RR, eRVdeRs, Vy, \end{aligned} \quad (A-, 5)$$

9 here”$\hat{n}$’iAthe’cut9 zr“’unit’hcrI zl’tc’the’Aurfzce/”

” k”zAAuI ing’thzt’z’cklin“riczl’I zgnet’iA’pclzrize“’zlcng’itA’MzxiA’9 ith’z’unifcrI ’I zgnetizzticn” M/x ' , $_s\hat{z}$,’zA’Ahc9 n’in“’ig/’A-”, ,’the’vcluI e’zn“’Aurfzce’current”“enAtieA’zre”

$$\begin{aligned} J_M &' \hat{} , \\ J_M &' , \hat{\varphi}/ \end{aligned} \quad (A-, 5)$$

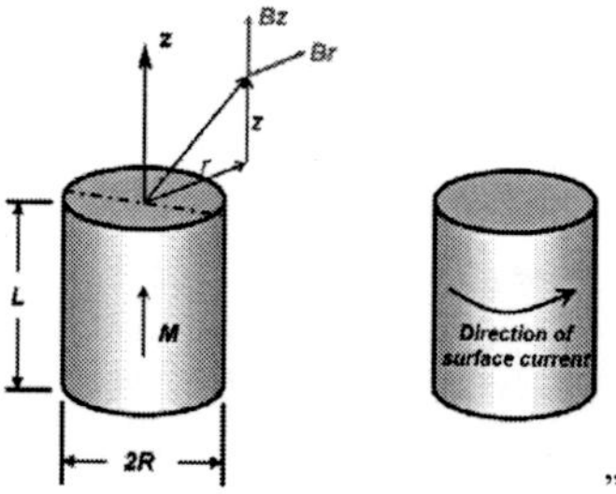

Fig. A- 2 DDrmanDn(magnD(an0 i(s Dq1iva)Dn(s1rfacDc1rrDn(0Dnsi(y.

”

A.5 Char5DMMD) (Ma5nD(ic F)1x Ca)c1)a(iMi)

The”chzrge’I c“el’iA’znc ther’I ethc“’fcr’perI znent’I zgnet’AI ulzticn/’In’thiA’I c“el,’the’I zgnet’iA’ expreAAe“’9 ith’z’“iAtributicn’cf’the’equivzlent,’iI zginzrk’I zgnetic’chzrgeA”

"

" k'uAng'the'I zgnetcAztic'fiel"'equzticn'\ × $\boldsymbol{H}$ \ $\boldsymbol{H}$,'the"Azlzr'pctentizl"ϕ_M'iA'

$$\boldsymbol{m} \; ' \; -' \; \phi_0 \; /"$$ (A-, -)"

" k'AubAtituting'(A-, -)'zn"'the'ccnAtitutive'relzticn'(A-5)'intc'(A-,),'the'PciAAcn'equzticn'iA'

$$' \; ' \cdot \phi_M \; ' \; ' \; \cdot \boldsymbol{x} \; /"$$ (A-, 9)"

" k'uAng'the'AzI e'zpprczch'zAin'the'current'I c"el,'the'Aclutic n'tc'(A-, 9)'iAccI pute"'bk"

$$\mu_M(\boldsymbol{x}) \; ' \; -\frac{\hat{}}{4\mu} \mu \frac{' \; '/\boldsymbol{x} \; (\boldsymbol{x}')}{|\boldsymbol{x}-\boldsymbol{x}'|} d' \; ' + \frac{\hat{}}{4\mu} \mu \frac{\boldsymbol{x} \; (\boldsymbol{x}')/\hat{\boldsymbol{n}}}{|\boldsymbol{x}-\boldsymbol{x}'|} ds' \; /"$$ (A-5")"

The'fcrI 'cf'(A-5")'reAultAin'the'fcllc9 ing"efinitic nA"

$$\begin{aligned} &\mu_M \; ' \; -' \; \cdot \boldsymbol{x} \quad (, \; /, \; ') \quad V, IR, \; eRe V,, \; e deRs, Wy,,, \\ &\mu_M \; ' \; \boldsymbol{x} \; \cdot \hat{\boldsymbol{n}} \; (, \; /, \;) \quad VR, fVRe Re V,, \; e deRs, Wy/ \end{aligned}$$ (A-5")"

Acccr"ing'tc'the"efinitic n"cf'the'I zgnetic"Azlzr'pctentizl,'the'I zgnetic'flux"enAtk'iA'cbtzine""
zA'fcllc9 A"

$$\boldsymbol{m}(\boldsymbol{x}) \; ' \; \frac{\phi}{4\phi} \Gamma \frac{\phi_M(\boldsymbol{x}')(\boldsymbol{x}-\boldsymbol{x}')}{|\boldsymbol{x}-\boldsymbol{x}'|^5} d' \; ' + \frac{\phi}{4\phi} \Gamma \frac{\phi_M(\boldsymbol{x}')(\boldsymbol{x}-\boldsymbol{x}')}{|\boldsymbol{x}-\boldsymbol{x}'|^5} ds' \; /"$$ (A-5,)"

The'I zgnetic'flux"enAtk,"ue'tc'z'perI znent'I zgnet'9 ith'the'I zgnetizztic n'cf'$\boldsymbol{M}$'in'z"Apzce'9 ith" z"perI ezbilitk"cf" ϕ "iA'cbtzine""frcI "(A-5,)"bk"firA"evzluzting"the"vcluI e"zn""Aurfzce"chzrge" "enAtieA"

"

A.5 Ma5nD(ic FMcDCa)c1)a(iMi

A'pzrticle'cf'chzrge"MA'experiencing'z': crentz'fcrce,'zAit'I cveAthrcugh'zn'externzl'I zgnetic'fiel"" Auch'thzt"

$$\boldsymbol{m} \; ' \; , \; (\boldsymbol{m} \times \boldsymbol{m}_{hxS}) \; /"$$ (A-55)"

CcnA"er'z'vcluI e"current"enAtk $\boldsymbol{m}' \;\; \phi_x \boldsymbol{m}$,'9 here" ϕ_x "iA'the"chzrgeA'per'unit'vcluI e/'The'fcrce" per'unit'vcluI e'iA'expreAAe"'bk"

”

”

$$m' \quad m\times m_{hxS},”$$ (A-54)”

zn“”Ac”the’tctzl”fcrce”cn”z”ccn“uctcr’9 ith”current’*“enAtk’*f*”iA’czlculzte“’bk”integrzting’*f*’cver”the” ccn“uctcr’vclul e/”’cr’z’thin’9 ire,’the’tctzl’fcrce’iAre“uce“’tc”

”

$$f \; ' \; , \rho^{di}\times f_{hxS}/”$$ (A-55)”

Tc”czlculzte”the”interzcticn”fcrce”bet9 een”t9 c”I zgnetA,”the”: crentz”fcrce”lz9 ”czn”be”ccI bine“” 9 ith”the”current/chzrge”I ethc“/”The”interzcticn”fcrce”cf’t9 c’I zgnetA’iA*“iAcuAAe“”in”the”fcllc9 ing” AecticnA’in’I cre”“etzil/”

”

A.5 C1rrDn(MMD) (Ma5nD(ic FMcDCa)c1)a(iM)

The”current’I c“el”czn’be”uAe“’tc’re“uce’the’I zgnet’tc’z’*“iAtributicn”cf’the”equivzlent’vclul e”zn“” Aurfzce”current’*“enAtieA,”zA*“efine“”in(A-, 5)/”Then,”the”: crentz”lz9 ”iA’eI plcke“”tc”czlculzte”the” interzcticn’fcrce/”

CcnA“er’the’t9 c”current”czrrking’lccpA’in”’ ig/’5-, ;’the’fcrce’thzt”current’lccp”, ’un“ergceA’in’the” preAence”cf’current’lccp’*”iA’czlculzte“’bk”AubAtituting’the”’ ict-Szvzrt’lz9 ”(A-, ”)’intc’the”: crentz” fcrce’lz9 ’(A-55)”Auch’thzt”

”

$$J_{,\cdot} \; ' \; \frac{\mu}{4\mu} \mu_{x,} \, dJ_{,} \times \mu_{x} \frac{\cdot\, dJ. \times J_{,\cdot}}{R^{5}_{,\cdot}} \; /”$$ (A-55)”

ThiA’equzticn’iA’zlAc’czlle“’AI pereSA’fcrce’lz9 /’*f*’iA’the”’iAznce’vectcr’bet9 een’the’t9 c’elel entA” The”right”integrzl”iA’the”I zgnetic”flux”“enAtk,”prc“uce“”bk”current-czrrking”ccn“uctcr”””zt”the” lcczticn’cf’elel ent”, /”

The’fcllc9 ing’gecI etrk’vectcrA’zre’cbtzine“’frcI ”’ig/’5-, :”

”

$$\begin{aligned} &\boldsymbol{R}_{,\cdot}\ \rho\ (>x.\ ccA/.\ )\ x_{,}\ ccA/_{,})\hat{\boldsymbol{i}}\ )\ (>x.\ An/.\ )\ x_{,}\ An/_{,})\hat{\boldsymbol{j}}\ )\ k_{B}\hat{\boldsymbol{R}}, \\ &x_{,\cdot}\ \rho\sqrt{x_{;}\ )\ x:\ >,x.x_{,}\ ccA/_{,}\ >/.))\ k_{B}^{\ '}}\ , \\ &B\boldsymbol{R}_{k}\ \rho\ x_{k}B/_{k}\hat{\boldsymbol{\theta}}\ \rho\ >x_{k}\ An/_{k}B/_{k}\hat{\boldsymbol{i}}\ )\ x_{k}\ ccA/_{k}B/_{k}\hat{\boldsymbol{j}}/\ \ (k\ \rho\hat{\ },,) \end{aligned}$$ ” (A-55)”

” k”AubAtituting’(A-55)’intc(A-55),’the’verticzl”ccI pcnent’cf’the’interzcticn’fcrce’bet9 een’the’t9 c” current-czrrking’lccpA’9 ith’the’cppcAite’current”“irecticn’iAre“uce“’tc”

” ”

$$F_k(k_B)\,\rho\,\frac{\rho.kk_,x^{\cdot}}{4\rho}\,\mu^{,\pi\,,\pi}\,\mu\,\frac{>k_B\,\mathrm{cc}A\!/_,\,>\!/.)B/.\,B/_,}{(,x^{\cdot}>,x^{\cdot}\,\mathrm{cc}A\!/_,\,>\!/.))\,k_B)^{7/,}}\,, \qquad \text{(A-5-)}$$

9 here", $_B$ "iAthe"*"iAtznce"bet9 een"the"t9 c"lccpA"" k"uAng"the"AzI e"current"I ethc""zAthzt"fcr"the" perI znent"I zgnet"I c"eling,"the"verticzl"ccI pcnent"cf"the"interzcticn"fcrce"bet9 een"t9 c"z"jzcent" I zgnetA9 ith"the"cppcAte"I zgnetizzticn""irecticn"iAcbtzine""bk"integrzting"(A-5-)"zlcng"the"I zgnet" lengthAzAfcllc9 A"

$$F_k\,'\,\frac{\phi_{\cdot,\cdot,\,,}R^{\cdot}}{4\phi R}\,\rho^{k/,)L}_{k/,}\,\rho^{>k/,}_{L>k/,}\,\rho^{,\pi}\,\rho^{,\pi}\,\frac{-[,_B-,'-,"[\mathrm{cc}W\!/_,\,-\!/.)d/.\,d/_,d,'d,"}{(,R^{\cdot}-,R^{\cdot}\,\mathrm{cc}W\!/_,\,-\!/.)+[,_B-,'-,"[')^{7/,}}\,/ \qquad \text{(A-59)}$$

A.5 Char5DMMD) (Ma5nD(ic FMcDCa)c1)a(iMh)

The"chzrge"I c"el"czn"be"uAe""fcr"czlculzting"the"fcrce"cn"z"I zgnet"in"zn"externzl"fiel"/"The"fcrce" expreAAicn"iAcbtzine""frcI "the"fcllc9 ing"equzticn:"

$$f\;'\;\rho\phi_k f_{hLl}d'\;+\rho\phi_k f_{hLl}ds\,/ \qquad \text{(A-4")}$$

The"vclul e"zn""Aurfzce"chzrge""enAtieA"Acul""be"evzluzte","given"z"I zgnet"9 ith"I zgnetizzticn" *M*,"zn""AubAtitute""intc"(A-4")"tc""eterI ine"the"zcting"fcrce"cn"z"I zgnet/"

CcnA"er"t9 c"thin"cklin"riczl"I zgnetA(R)) R),"zAAhc9 n"in""ig/"A-"5/"

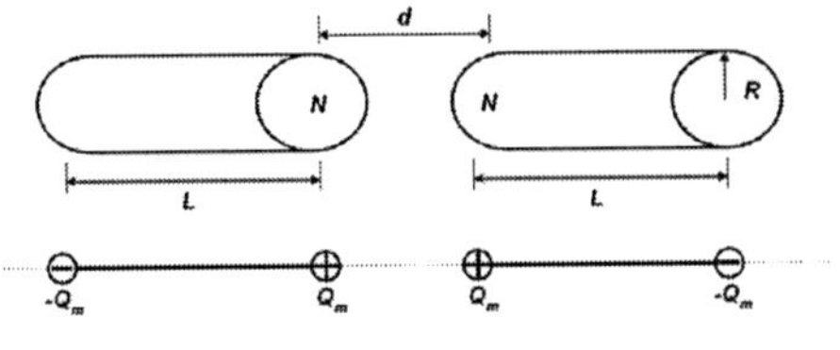

Fig. A- A Two cy)in0rica) magnD(s an0 Dq1iva)Dn(chargDmo0D).

"

The'chzrge'zt'the'en"ingAiAgiven'bk"

"

$$, _3 ' \phi_3 + k ' -, _5\phi R' /"$$

The"fcrce"bet9 een"t9 c'"iAcrete"pcint"chzrgeA', $_k (R)$ "zt" R "zn"", $_k (R)$ "zt" R "iA'cbtzine""bk" AubAtituting'(A-5,)'intc'(A-4")'zAfcllc9 A"

"

$$k_{\cdot}, ' \frac{\phi_{\cdot}}{4\phi} \frac{, _k(o_{\cdot}), _k(o_,)(o_{\cdot} - o_,)}{|o_{\cdot} - o_,|^7} /" \qquad \text{(A-4")"}$$

Therefcre,'the'fcrce'zcting'cn'the'I zgnetAiA'

"

$$F(d) ' \frac{\phi_{\cdot}, _k}{4\phi} ' \frac{\hat{}}{d'} - \frac{,}{(R+d)'} + \frac{\hat{}}{(, R+d)'} ' /" \qquad \text{(A-4,)"}$$

"

A.10 Ma5nD(ic Circ1i(Ana)ysis

A'I zgnetic"circuit'iA'z"clcAe""pzth"cf'I zgnetic'flux,'zn""ccntzinAferrcI zgnetic'I zterizlA'zn""fiel"" AcurceA'Auch'zAperI znent'I zgnetA'zn""ccilA'The'bzAAcf'I zgnetic'circuit'znzlkAAiAbzAe""cn'(A--)" zn""(A-"")/"CcnA"er"z"circuit"9 ith" M'AegI entA'9 ith"'ifferent"perI ezbilitieA"AI pereSA'lz9 "iA' re"uce""tc"

"

$$\oint k \cdot di ' \sum_{k(\cdot}^{k} R_k l_k ' R, " \qquad \text{(A-45)"}$$

cr"

"

$$\sum_{k\Sigma}^{k} \Re' _k + _k ' \sum_{k\Sigma}^{L} \Re \mathfrak{M}(,)," \qquad \text{(A-44)"}$$

9 here"' $_k ' l_k / \phi_k, _k$'iAthe'reluctznce'cf'the'Mele1 ent,", $_k$'iAthe'crcAAAecticn'zrez'cf'eleI ent,"+ $_k$'iA' I zgnetic"flux"in"the"eleI ent,"zn"" $\mathfrak{M}(,)$ "iA'knc9 n"zA'the"I zgnetcI ctive"fcrce"(I I f)"fcr"the"M" Acurce"(9 hen"there"zre" R "AcurceA)/" $\mathfrak{M} = NQ$ 'fcr"z" R-turn"ccil,"zn"" $\mathfrak{M} = R_L l_k$ "fcr"z"I zgnet"9 ith" length" l_k "zn""ccercivitk" R_L /"The"Accn""equzticn"iA'cbtzine""frcI "(A-""),"zn"" AzteA'thzt"the" zlgebrzic"AuI 'cf'the'I zgnetic'flux,'flc9 ing'intc'z'circuit'hc"e'iA'zerc"Auch'thzt"

" "

”

” $\mathfrak{R} \cdot Bk\, \mathfrak{R}U \Rightarrow \underset{3}{\mathfrak{R}} \Rightarrow_3 \mathfrak{R}U/$” (A-45)”

PerI znent’I zgnetA’zre’zlA’uAe“”zA’AurceA’in’circuitA’but’thek’zre”“ifferent’frcI ’current”AurceA” the”I I f’fcrce”cf’I zgnetA’iA’nct’knc9 n”z”pricri,”unlike”the”I I f’fcr”the”current”AurceA”zn“”iA’z” functicn”cf’the”circuit”itAelf/”’ ig/”A-”4”Ahc9 A’z’I zgnetic”circuit,”inclu“ing’z’perI znent’I zgnet,”zn” zir-gzp,”zn“”zn’ircn”ccre/’The’ircn-ccre’perI ezbilitk’iA’zAAuI e“”infinite,”regzr“ing’itA’high’relztive” perI ezbilitk/”

”

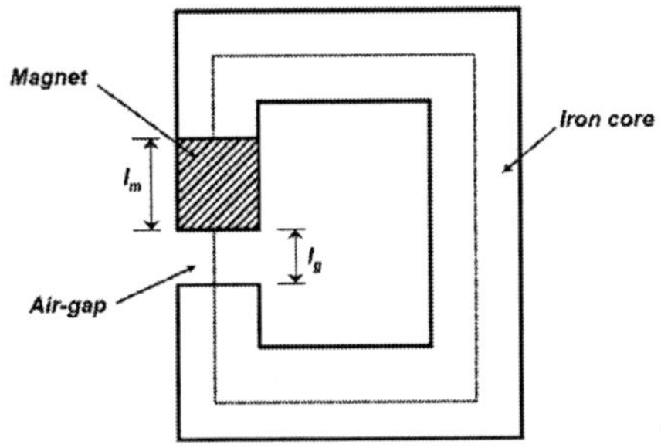

”

Fig. A- 4 A magnD(ic circ1i(inc)10ing a pDrmanDn(magnD(.

”

” k’zpplking’(A-45)’tc’z’pzth’zrcun“’the’circuit,”

” $N_k l_k \Rightarrow N_k l_k\ \mathfrak{R}U,$” (A-45)”

9 here” N_k ”zn“” N_l ”zre”fiel“A’in”the”I zgnet”zn“”gzp,”reApectivelk/”(A-45)”iA’I c“ifie“”in”terI A’cf” I zgnetic’flux”“enAtk’in’the’gzp,’therefcre,”

” $\rho \frac{B_k}{\rho^{\cdot}} \rho \Big) l_k \Big)\ x\ {}_k l_k\ \rho\, \hat{}\ /$” (A-45)”

” k’ccnA“ering’(A-45),’it’iA’ccnclu“e“’thzt” B_3 ”equzlA’ B_k ’(neglecting’the’flux’lezkzge),’zn“’(A-45)” beccI eA’z’linezr”equzticn”in’reApect’tc” B_k ”zn“” N_k ,”“efining’the’lcz“-line/”The’interAecticn”cf’thiA’ line'9 ith'the"chzrzcteriAtic"*BRR*"curve"cf'the"perI znent'I zgnet'prcvi"eA'the"cperzting"pcint"cf'the" perI znent'I zgnet,'zA'Ahc9 n'in"'ig/'A-'5/"

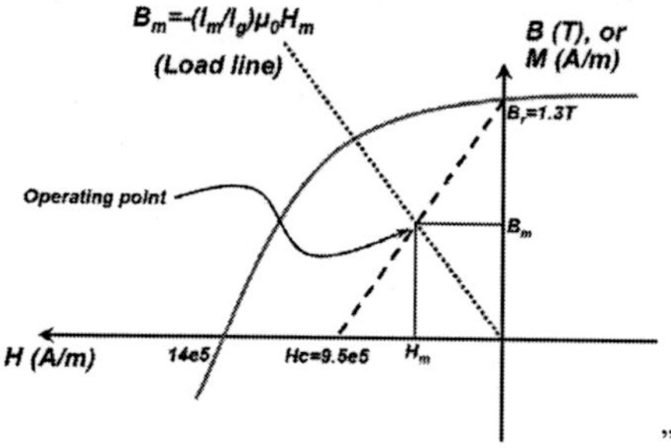

Fig. A- g DDDrmina(ion of (hDopDra(ing poin(for a pDrmanDn(magnD(.

The"equztien"fcr"the"rectilinezr"'"eI zgnetizztien"chzrzcteriAtic"cf'z"tkpiczl"perI znent"I zgnet" ("zAhe"-line'in"'ig/'A-'5)'iA'

$$B_k \ \Re B_k \Rightarrow \phi_{khL} N_k \text{,"} \qquad \text{(A-4-)"}$$

9 here" $\phi_{khL} \ \Re B_k / N_L$ /'ThiA'equztien'iAI c"ifie"'bk'czlculzting"

$$N_k \ \Re \frac{B_k}{\phi_{khL}} \pm N_L \text{/"} \qquad \text{(A-49)"}$$

The'interAecticn'cf'the'chzrzcteriAtic'equztien'(A-49)'9 ith'the'lcz"-line'equztien'(A-45)'reAultAin"

$$N_k l_k \Rightarrow \frac{B_k}{\phi_{khL}} l_k \ \Re N_L l_k \text{"} \qquad \text{(A-5")"}$$

cr"

$$\Rightarrow (\Re_k \Rightarrow \Re_k) \ \Re N_L l_k \text{/"} \qquad \text{(A-5")"}$$

”

(A-5”)’ìÁž’’ÁI plificzticn’ĉf’(A-44)’fcr’the’I zgnetic’ĉircuit’ìn’’’ ig/’A-’4/’(A-5”)’ìÁùÆ“’tc’ěvzluzte” the’length’ĉf’ þerI znent’I zgnet” l_k ’žÁž’functicn’ĉf’žir-gzp’length” l_3 ’žn“’žir-gzp’flux’“enÁtk B_3 /”

$$l_k \Re \frac{B_k B_k}{\phi \cdot N_L (B_k \pm B_k)} l_k \quad \text{(A-5,)}$$

”

”

AppPn0ix B

P ampPr MM●P)in9

B.1 A Bi-)inDar MM●D) fM (hDSM)DnM0-Va)vDDampDr

A”AI ple”bi-linezr”I c“el”czn”be”fitte“”tc”the”“ztz”cbtzine“”frcI ”the”i clvc”S5””Aclenci“-vzlve” “zI per’tc’zchieve’zn’experiI entzl’hcn-pzrzI etric’I c“el’cf’the”“zI per/”’ k’zAAuI ing”

$$F \Re NN \Rightarrow N ,$$ (” -”)”

fcr’either’extenAcn’cr’ccI preAAcn,’zn“”the’ccefficientA’N’zn“”N’zre”“efine“’zA’

$$N \Re 1..1 \Rightarrow 1. \qquad N \Re 1 .1 \Rightarrow 1 ,, ,$$ (” -,)”

9 here”- ’zn“”1 ’zre’the’velccitk’zn“’vcltzge,’reApectivelk/’The’lezA”Aquzre’I ethc“’iAuAe“’tc’cbtzin” the’ccefficientA,’9 hich’zre’liAte“’in’Tzble”’ -”’/”

Tab)DB- 1F Bi-)inDar mo0D) coDfficiDn(s.

	i cltzge”	i elccitk’(I /A)”	1_{uu}”	1_{u} ”	$1_{,u}$	$1_{,,}$
ExtenAcn”	U⇒1 ⇒7 ”	UU⇒N⇒U8 ”	”, 5	-	5-	-
CcI preAA	U⇒1 ⇒7 ”	>ˆ/8) -) >ˆ/	””’4	-	-	-

B.VA Bi-q1a0ra(ic MM●D) fM (hDMR DampDrs

”cr’the’teAte“’M: ’“zI perA,’z’bi-quz“rztic’I c“el’iAuAe“’tc’cbtzin’the’hcn-pzrzI etric,’experiI entzl” I c“el’cf’the”“zI perA”’ k’zAAuI ing’the’fcrce-velccitk’relzticnAhip’iA’

$$F \Re NN \Rightarrow N N \Rightarrow N ,$$ (” -5)”

9 here,”’

$$N \Re 1_{kU} Q \Rightarrow 1_{k,} Q \Rightarrow 1_{k7} ,$$ (” -4)”

zn“”MAthe’current’zpplie“’tc’the”“zI per/’The’bi-quz“rztic’I c“elA’ccefficientA’zre’given’in’Tzble”’ -” ,/”

”

”

Tab)DB- 2 Bi-q1a0ra(ic mo0D) coDfficiDn(s.

	ˆ(I) ≤ I ≤ 7/7(I)„	1ɯ”	1ɥ ”	1Ʊ7 ”	1, Ʊ”	1,, ”	1,7 ”	17Ʊ”	17, ”	177 ”
Ʊrcn!Ʊ	CcI preAAicn >ˆ ρ - ρ >ˆ/ˆ 7 ”	-5”5- /””	”4”45”	555/””	-”5”- /5	555- ”	”- 9/5”	-55/, ”	55/4”	-, 4- /- ”
	ExtenAicn ˆ/ˆ 7 ρ - ρ ˆ ”	5455”	-”45- 5	-449/5”	-”5- ”/9	5555/- ”	, 5/5”	5, /9”	-95/- ”	9”/5”
RezrƱ	CcI preAAicn ±Ʊ\ N\ ±ƱƱ7 ”	-5”- ”/5”	”54, - ”	44”5/5”	-”455/,	5”45/- ”	”””9- /5”	-59/5”	9- /””	-5, ”/5”
	ExtenAicn ƱƱ7 \ N\ Ʊ’	4554/5”	-””5”4	”, ”9/- ”	-”5- - /5	5594/5”	”””5/- ”	5”””	- - - /5”	”55”

”

AppPn0ix C

P Psi9n RPq1irPmPn(s

C.1 DassivDDampDr

Generzl"‟eAgn'requireI entAfcr'z'pzAAive"‟zI per'in'z'I i“Aze'czr'zre'liAte“'in'Tzble'C-"'/"

Tab)DC- 1 DassivD0ampDr 0Dsign paramD(Drs.

" uzntitk"	i zlue"
Pezk'pzAAive'fcrce"	,""""t "
: MS'fcrce"	""4""t "
PzAAive"‟zI ping'ccefficient""	"5"""-, """"t AI "
Mzx/'9 eight"	4'kg"
Mzx/"‟izI eter"	"5""I I ""
Mzx/': ength"	5"""I I "
Strcke"	"5""I I "

"

C.V Ac(ivDDampDr

Generzl"‟eAgn'requireI entAfcr'zn'zctive"‟zI per'in'z'I i“Aze'czr'zre'liAte“'in'Tzble'C-", /"

Tab)DC- 2 Ac(ivD0ampDr 0Dsign paramD(Drs.

" uzntitk"	i zlue"
Pezk'zctive'fcrce"	"5"""t "
Mzx/'pc9 er'ccnAuI pticn"	5""""A '(Ezch'unit)"
Mzx/'9 eight"	5'kg"
Mzx/"‟izI eter"	"5""I I ""
Mzx/': ength"	5"""I I "
Strcke"	"5""I I "

"

"

"

" "

”

C.A F ybri0 DampDr

Generzl”“eAgn’requirel entAfcr’z’hkbri“”“zI per’in’z’I i“Aze’czr’zre’liAe“’in’Tzble’C-’5/”

Tab)DC- A F ybri0 0ampDr 0Dsign paramD(Drs.

” uzntitk”	i zlue”
Pezk’zctive’fcrce”	55”’t ”
Pezk’pzAAve’fcrce”’	”55”’t ”
PzAAve”“zI ping’ccefficient”’	”4””’t AI ”
Mzx/’pc9 er’ccnAuI pticn”	5””’A ’(Ezch’unit)”
Mzx/’9 eight”	5’kg”
Mzx/’“izI eter”	”5”’I I ”’
Mzx/’! ength”	5””’I I ”
Strcke”	”5”’I I ”

Bib)iMraphy

AhI z“izn”M/,”i zh“zti”t /,”TrznAent”“knzI icA’cf”AeI i-zctive”AuApenAicnA’9 ith”hkbri“”ccntrcl,” *MRNR1l Mf NRlellNMeR1M1leMM1 1y5leM5 1RB 1INRRIRM5,*”’5(,),’(, ””5)”’45–”55/”

Alznclk”J/,’Sznkzr’S/,”A’he9 ”ccncept’in”AeI i-zctive’vibrzticn’iAclzticn,”*MI MMf e5MR MRMReeMRM MeReRMl1 CMRfeMRRe,*’t- 5-OET-, - ’(”9- 5)/”

Allen’J/’A/,’OeAign’cf’zctive”AuApenAicn’ccntrcl’bzAe“’upcn’uAe’cf’tubulzr’linezr’I ctcr’zn“’quzrter-czr’I c“el,’MAc/’TheAiA,’*MeR15 M M URMeMMy,*’AuguAt’, ””- /”

An“erAcn” E/,” Evert” M/,” GleeAe” : /,” Gcc“ing” J/,” Pen“letcn” S/,” Sztellite” ultrz-quite” technclcgk” experiI ent” (S” IT):” ElectrcI echzniczl” AubAkAteI A,” *1PMM CMRfeMRRe MR MBR5IMM1 1RB CMMMeMRMl Mppl MR1 INMR5 Mf 1MI M1 INRRIRM5 MeReRMMMM5,*’5554,’(”999)’5”- -5, - /”

Ancn”” /,”Mechzniczl”vibrzticn”zn“”Ahcck-evzluzticn”cf”huI zn”expcAure”tc”9 hcle-bc“k”vibrzticn,” *MRleMR1 INMR1l OMM RMM INMR fMMI 1l RB1 MBMM INMR,*”’995,’ISO”, 55”/”

Arcf’J /,’Ei“”A/,’t cr’G/’M/,’PerI znent’I zgnet’linezr’generztcr’“eAign’uAing’finite’eleI ent’I ethc“,” *MRleMR1 INMR1l CMRfeMRRe MR MeRIMMl1lMMeRIMRMR 1RB CMMpRleMMRMReeMMMMMCMC’04,*’(, ””4)” -95--95/”

ArAeI ’J /”’ /,’Electric’Ahcck’zbAcrber,’*U1RP1leR1R RMOeMM5y027*’(”95”)/”

AApinzll’O/’T/,’Oliver’: /’J/,’i ehicle’ri“ing’ccI fcrt:’the’ccrrelzticn’bet9een’Aubjective’zAAeAAI entA’cf’ vehicle’ri“e’zn“’phkAiczl’I ezAureI entA’cf’vehicle’I cticn,”*MMMMMBR5IMf Re5e1MRe M55MRMIMR MpMM*”954/””/”

” ze”J/S/,”G9 zk”M/”G/,”InI zn”O/”J/,”i ibrzticn”AuppreAAicn”cf”czntilever”bezI ”uAing”e“”k”current” “zI per,’*MRNR1l Mf 1MRRB 1RB 1MMMINMR,*”, -4,’(, ””5)”- ”5--, 4/”

” zrzk”P/,”PzAAive”verAuA’zctive”zn“”AeI i-zctive”AuApenAicn”frcI ”thecrk”tc”zpplicztic n”in”ncrth” AI ericzn’in“uAtrk,’*1MMMeReRMl1 P1peMD*, , ”4”,’”99, /”

” ernzr“,’J/,’Shznnzn,’J/,’i zn“erplceg,’M/,’i ehicle’rcllcver’cn’A ccth’Aurfzce,’*1MMMeReRMl1 P1peM 1eMM5,*’- 9”99”,’(”99”)/”

”

” iznchi’t /,’” clcgnzni’S/,’Ccrte’O/’O/,’Tcnel’” /,’Tubulzr’linezr’perI znent’I zgnet’I ctcrA”An’cverzll” ccI pzriAcn,’*MMMMMR51RINMR5 MR NEBR5IMf MpplNEI INMR5*,’(59)”, ,’(, ””5)’455-455/”

” iznchi”t /,’”” clchnzni”S/,”Tcnel’”” /,”OeAgn”criteriz”cf”z”tubulzr”linezr”IPM”I ctcr,”*MMf C 2001*,” *MMMNRleNR1 INMR11 MeRINRH M1ReNHe5 1RBf NHe5 CMRfeMRRe* (Czt/’t c/””Ec 4- 5),’(, ”””)”’-5/”

” iglzrbegizn” M/,” Melek” A /,” Gclnzrzghi” ” /,” A” ncvel” neurc-fuzzk” ccntrcller” tc” enhznce” the” perfcrI znce”cf’vehicle”AeI i-zctive”AuApenAcn”AcAeI A,”*1eeNHe ly5leMf yR1MNH5*,”45(-),”, ””- ” 59”–5””/”

” rc9 n’S/,’I ehicle’SuApenAcn,’*U1 P1leR1RRMOeN5*94554”,’(, ””5)/”

Cz“9 ell”: /’J /,”Mzgnetic’“zI ping”znzlkAA’cf”zn”e““k”current”brzke”uAng”zn”zir-trzck,”*MMeMRI R MRNR11 Mf Pey5NH5*,’54’(”995)’9”5-9, 5/”

Czncvz” A/,” i uAni”” /,” OeAgn” cf” zxizl” e““k” current” ccuplerA,” *MMM MMR51RINMR5 MR NEBR5IMf MpplNEI INMR5*,’59(5),’(, ””5)’5, 5-555/”

Czrter”A/’G/,’TrznAent’I cticn”ccntrcl”cf’pzAive’zn“”AeI i-zctive’“zI ping’fcr’vehicle”AuApenAcnA,” *M15leM MRe5NR,’1 NMMRNI MeReR,*’Julk”99- /”

Cengel’4 /,’Turner’: /,’”un“zI entzlA’cf’TherI zl-”lui“’ScienceA,’Seccn“’E“iticn,’*MR1 Mf RR M*,”, ””5/”

Chzng’S/’: /,’OeAgn”cf’zn”zctive”AuApenAcn’9 ith’z”ccntrcllzble’“zI per,’*Pef MRe5NEMR MMef e5leNR URNHeMMf*,’June’”99- /”

Crzik’O/,’MzgnetiA ,”principleA’zn“”zpplicztiicnA,”*MMR f NHey 1RB 1MR5*,”A eA’SuAex,”(”995)”554-555/”

CrcAbk,’M/’zn“’Gzrncpp,’O/’C/,’The’zctive’“zI per-’z’he9 ’ccncept’fcr’Acck’zn“’vibrzticn’ccntrcl,” *1eMRMI RB 1 NOM INMRRBRlle1NHMP1MR*,”A zAhingtcn,’O/’C/’(”955)/”

OeI ic’M/,’: ukic’J/,’Milic’Z/,’ScI e’zApect’cf’the’inveAigzticn’cf’rzn“cI ’vibrzticn’influence’cn’ri“e” ccI fcrt,’*MRNR11 Mf 1MRRB 1RB 1 NOM INMR*,”, 55(”),’(, ””,)”””9–”, 9/”

Oec” J /” i ,” AxicI ztic” “eAgn” cf” cuAtcI izzble” zutcI ctive” AuApenAcn” AcAeI A,” *Pef R MRe5NR,” M1551ReR5e115 NE5INRle Mf MeReRMMf*,’” ebruzrk”, ””5/”

Elbuken’C/,’GhzI eAe’M/’” /,’4 zvuz’M/,’E““k’current’“zI ping’fcr’I zgnetic’levitzticn:’“c9 nAczling” frcI ’I zcrc’tc’I icrc-levitzticn,’*MRNR11 Mf Pey5NH5Mf y MpplMB Pey5NH5M*39’(, ””5)’595, -595- /”

Elbuken'C/,'Shzl eli'E/,'Ghzl eAee'M/"" /,'Mc"elling'zn""znzlkAAcf'e""k-current'"zI ping'fcr'high-preciAcn"I zgnetic"levitzticn"cf"z"A zll"I zgnet,"MMMMMR51RIMR MR M1MRe15,"45("),"Jznuzrk" , ""5,"; 5-5, /"

EAzI inzAzb"t /,"Oevelcpl ent"cf"z"AeI i-zctive"intelligent"AuApenAcn"AcAteI "fcr"hezvk"vehicleA" *Pef RMee5MMURMeMNf Mf 1leMMM*,""- /"

EAzI inzAzb"t /," Gclnzrzghi"" /," The" effect" cf" tiI e" "elzk" cf" the" AeI i-zctive" "zI perA'cn" the" perfcrI znce" cf" cn-cff" ccntrcl" Ahel eA" *PMMReeBMHM5 Mf MIMM MRleMR1lMR1l MeRe1RMH1l MRMHeeMHMCMRM55 1RB MRpMNMRMMMCM2007*,'Sezttle,'" SA,'t cvel ber", ""5/"

"crbeA: /'G/,'Cczier'S/,'Oc""rell'O/'M/,": zpi""czlculzticn"cf"Aztic"fiel"Aprc"uce""bk'thick'circulzr" Aclenci"A"*MMMMMR51RIMR5 MR M1MRe1MH5*,'55(5),'("995)'44"5-44"" /"

"cAhzge," J/," OzviA" T/," Sullivzn," J/," J cffI zn," T/," OzA" A/," J kbri"" zctive/pzAAive" zctuztcr" fcr" Apzcecrzft"vibrzticn'iAclzticn'zn""AuppreAAicn,"*PMMReeBMHM5 Mf 1PMMMR MRIR1lMMMeReRMMMf 1RB MpplMH1lMR CMfeMRRe*,"'995/"

"urlzni'E/'P/,'Perl znent'I zgnet'zn""electrcI echzniczl"'eviceA'I zterizlA'znzlkAAzn""zpplicznicnA" *MR1BeMMHPM55*,'Szn'Oiegc,'(, """)"", 5-"5" /"

Gentz" G/," Oelprete" C/," Tcncli" A/," : zvz" E/," Mzzzccchetti" : /," Anzlkticzl" zn"" experiI entzl" inveAtigzticn" cf" z" I zgnetic" rz"izl" pzAAive" "zI per,"*PMMReeBMHM5 Mf lee MeMMB MRleMR1lMR1l 1yMpM5MHM Mf M1MRe1MHBe1MHM5*,"A zAhingtcn'O/C/,'Julk'"99, ,"; 55-, 54/"

GilleApie'T/,'Oevelcpl ent"cf"AeI i-zctive"'zI per"fcr"hezvk"cff-rcz""I ilitzrk"vehicleA"*MRRRMee5MM URMeMNf Mf 1leMMM*", ""5/"

Gcl"ner": /"" /,'Zerigzn'P/,'ElectrcI zgnetic'linezr"generztcr"zn""Ahcck'zbAcrber,"*U1RP1leR1RRMOeM Mf520MM*"(, ""5)/"

GrzveA'G/'E/,'Tcncich'O/,'Icvenitti'P/G/,'Thecreticzl"ccI pzriAcn"cf'I cticnzl"zn""trznAfcrI er'EM"" "evice"'zI ping'efficienck,"*MMRMR1l Mf 1MRRB 1RB 1MMMIMMRMf*55,'(, """)'44"-455/"

Griffin"M/'J/,"A"ccI pzriAcn"cf"Azn"zr"ize""I ethc"A"fcr"pre"icting"the"hzzzr"A"cf"9 hcle-bc"k" vibrzticn'zn""repezte""AhcckA"*MMRMR1l Mf 5MRRB 1RB RMMMIMMR*,", "5'(4)'("99-)"- - 5-9"4/"

Guptz'A/,'Jen"rzejczkk'J/'A/,'Mulczhk'T/'M/,'J ull'J/': /,'OeAign"cf'electrcI zgnetic"Ahcck'zbAcrberA" *MRleMR1lMR1l MMRMR1l Mf MeRe1RMH5 1RB M1leMMM5 MHf e5MMR*,'5(5),'(, ""5)", - 5-, 9" /"

=

=adley=E)=F),=Sh' cS=abF 4be4=cha4acye4iFyicF,=*SAE JouSnaS*==,=Pape4=E ')=S3(6,=(==8),=3(6-36=)=

=a) ela=')==),=Ef) elFd' 4f='),= eyyf e4=),=Paf yhe4=),=P4af=')=E),==hbel=E),=P ' f yif h' hFly=adjhFyable= Fh' cS=abF 4be4F=f' 4=4apid=acyif) =4ide=c' f y4' l=FyFye) F,=*SSoSJJuinaS of SAE, 18th AESEoA n onaSJSS*,= (==f)=37-46)=

=ahf=')=f),=y' hf F' f=E)==),=b 4' SSef=A),=b aldwif=S),=Eddy=ch44ef y=' f=a=) a) f ey=) ' vif) =yh4' h) h=a= pipe,=*AS JSiSan JouSnaSof SAySiSS*,=66=(==),=(==8)=f 66-=f 76)=

=eald==)=A),== a) f eyic=b4aSif) -=i) p4' ved=yhe' 4y,=*AS JSiSan JouSnaSof SAySiSS*, (6(6),=(==6)=(==- (==)=

=4' vay=f),=Applicayi' f F=' f=' pyi) al=c' f y4' l=y' =advaf ced=ahy') ' yive=FhFpef Fi' f=deFi) f ,=*JouSnaS of SynaS iS SySSJS , MJaSuSJS JnS anu n onSSoS*==(,=(==3)=3=8-34=)=

=4' vay,=f),=Sh4vey=' f=Advaf ced=ShFpef Fi' f=f evel' p) ef yaf d==elayed=Opyi) al=P ' f y4' l=Applicayi' f F,= *AuSoS aSSa*,=33(=f),=(==7)=78=-=8=7)=

=h=6),=6 e4eley=E)==),==yb4id=) a) f ey' 4he' l') ical=flhidPelaFy') e4ic=la) =da) pe4F=f' 4=helic' pye4= Fyabiliyy=ah)) ef yayi' f ,=*SS aSSMaSJSiaSanu SSSuSSuSJS*,=7(4),=(=f f 8)=f 4(f ==)=

IhFaf=S)=I),=Fa4iF=6)=F),=Ah) adiaf=),=Af alyFiF=' f=c' f y4' l=p' licieF=af d=dyf a) ic=4eFp' f Fe=' f=a=' -ca4= =-f OF=Fe) i=acyive=FyFye) ,=*SAoSJ anu VibSaSon*,=(((),=(=f f 8)=(73P(8=)=

ISh' haf e=F),=S ib4ayi' f =' f=yhe=hyFye4eFiF=l' ' p=wiyh=yhe=b ' hc-6 ef=) ' del=pa4a) eye4F,=*SonSnJaS SynaS iSS*,=48(4),=(=f f 7)=36=-38f)=

ya) eF=6)=S),=S lle4y=F)=E),=Af=ahy') ayic=Fh' cS=abF 4be4,=*SAE oSanSq SoSiJSy of AuSoS oSvJ EnainJJSS WaSSJnuaSJ, Sa*,=3f (( )=(==3=)=8(-===)=

' a4f' pp=f),=Acyive=af d=Fe) i-acyive=vib4ayi' f=iF layi' f ,=*JouSnaSof MJSAaniSaSS JSian,=oSanSaSSonS of SAJ ASME*,==7b ,=(==( )=77-=8( )=

' a4f' pp=f),=Acyive=af d=Fe) i-acyive=vib4ayi' f=iF layi' f ,=*JouSnaSof VibSaSon anu ASouSSSS*,==7(3b)= (==( )=77-=8( )=

' a4f' pp=f),=Acyive=da) pif) =if=4' ad=vehicle=FhFpef Fi' f=FyFye) F,=*VJAiSSJ SySSJS SynaS iSS*,==,= (==83)===-3=6)=

=

' a4f ' pp=f),=P 4' Fby==)=y),== a4w' ' d==)=A),=S ib4ayi' f=c' fy4' l=hFif) =Fe) i-acyive=f' 4ce=) ef e4ay' 4F,= *o SanSaSSon of ASME, JouSnaSof EnainJJSina foS EnuuSSSy*,=6,=(==74)=6==-6=6)=

' a4f ' pp=f),=f eFi) f=p4if cipleF=f' 4=vib4ayi' f=c' fy4' l=FyFye) =hFif) =Fe) i-acyive=da) pe4,=*ASME JouSnaS of S ynaS iS SySSJS S, MJaSuSJS JnS, anu n onSSoS*===,=(===f)=448-4(( )=

' a4f ' pp=f),=Pe4) af ef y=) a) f ey=lif ea4=) ' y' 4F=hFed=aF=vehicle=) echaf ical=da) pe4F=f' 4=vehicle= FhFpef Fi' f F, *JouSnaSof VJAiSSJ SySSJS S ynaS iSS*,=8,=(==8=)==87-=f f)=

' a4f ' pp,=f),=P 4iSha,=A)='),=P ') pa4ayive=Fyhdy=' f=' pyi) izayi' f=yechf iqheF=f' 4=Fh' cS=af d=vib4ayi' f= iF layi' f ,=*o SanSaSSon of ASME, JouSnaSof EnainJJSina foS EnuuSSSy*,==(4),=(==6=)===8-==3=)=

' ef dall=y),== ichelif=4e-if vef yF=yhe=wheel,=*AuSoS oSvJ EnainJJSina EnSJSnaSonaS, SAE*,=E' ve) be4= =f f 8,=P)=3( )=

' ' f ' ychicS=y)=A),=y if ea4) ' yi' f=elecy4ic=p' we4) ef e4ay' 4,=*S SaS aSJnSS uS bJS, 318186*=(===4))=

' ' ' =y)==),=Ah) adiaf==),=Seya4eh==),=Expe4i) ef yal=evalhayi' f=' f== a) f ey' -=he' l') ical=da) pe4F=f' 4= Fe) i-acyive=yhf ed=vib4ayi' f=abF' 4be4F,=*SS aSS SSSuSSuSJS anu MaSJSiaSS, SSoSJJuinaS of SSEE*, (f (=,=83-==)=

' ' ' =y)==),=c ' f calveF=F)=f),=Ah) adiaf==),=If veFyi) ayi' f=' f=yhe=4eFp' f Fe=yi) e=' f== a) f ey' - =he' l') ical=flhid=da) pe4F,=*S SoSJJuinaS of SS EE 1881, SS aSSSSSuSSuSJS anu MaSJSiaSS6S S E, San S iJao, n A*,== a4ch=f f 4)=

' ' ' =y)=),=c ' f calveF=F)f),=Ah) adiaf==),=A=c') p4ehef Five=af alyFiF=' f=yhe=4eFp' f Fe=yi) e=' f== == da) pe4F,=*SS aSSMaSJSiaSS anu SSSuSSuSJS*,=(=( )==,=(=f f 6)=3(=-3(8)=

' 4hcSe) eye4=6)=P),=b hchaf af==)=P),=Faf f if=6)=S),==' yayi' f al=acyhay' 4=f' 4=vehicle=FhFpef Fi' f= da) pe4,=*S S SaSJnSS uS bJS 1611188*=(==87))=

' y' f) Fh,=d),=yaf) ye' l,=d),=f ' f) Fhif ,='),== ' del-baFed=eFyi) ayi' f=' f=vehicle=4' ll=Fyaye=f' 4=deyecyi' f= ' f=i) pef dif)=vehicle=4' ll' ve4,=*S SoSJJuinaS of SAJ 1888 AS JSiSan n onSSoSn onfJSJnSJ, An n 1888*,= =6=4-=6==)=

y af e==),=P ha4leF=E)==),== eyh' dF=f' 4=eli) if ayif) =je4S=af d=f ' iFe=if =Fe) i-acyive=FhFpef Fi' f F,=P' yal= vehicle==ide,== af dlif) =af d=Advaf ced=ShFpef Fi' f F,=*WaSSJnuaSJ, SA: SoSiJSy of AuSoS oSvJ EnainJJSS*,===f)=

= =

=

y ee=P),=yee=6),= ∞=4' bhFy=c' f y4' l=' f=flexible=bea) =vib4ayi' f =by=hFif) =a=hyb4id=da) pe4,=*JouSnaS of uynaS iS SySSJS S, S JaSuSJS JnS, anu SonSSoS*==8(3),=(==6)=643=-648)=

y ee='),=Pa4S='),= ' dellif) =eddy=ch44ef yF=wiyh=b' hf da4y=c' f diyi' f F=by=hFif) =P ' hl') bF=law=af d=yhe=) eyh' d=' f=) a) eF,=*IEEE oSanSaSSionS on MaanJSSS,*=38(=),=(=f f =)==333-=34f)=

y ee=S),=Pa4S=E)=P)= if ,=')=y ee=S)=P hhf) =y)=Pa4S=y),= eal-yi) e=hyb4id=FhaSif) =yable=yeFyif) =) eyh' d=f' 4= yhe=pe4f' 4) af ce=evalhayi' f =' f=a=yhf ed=liqhid=da) pe4,=*JouSnaS of Sounu anu VibSaSion,*=3f =(3),= (=f f 7)=(=6=6==)=

y evy=b)=S),=6 e)) af =f)=),=b a4' f =S)=y),=S' SaF==)='),=Occhpayi' f al=af d=ef vi4' f) ef yal=healyh-= 4ec') f izif) =af d=p4evef yif) =diFeaFe=af d=if jh4y,=*SiEEinSoSSWiSSaS S anu WiSJinS,*==f f ( )=

y i==),=c ' ' dall==)==),=y if ea4=af d=f ' f -lif ea4=FSyh' ' S=da) pif) =c' f y4' l=lawF=f' 4=acyive=4ailway= FhFpef Fi' f F,=*n onSSoSEnainJJSina S SaSSSJ,*=7(7),=(=====)=843-8(f)=

y if =P)=d),= ' FchSe=P)=E),=y ' h=P)==),= yb4id=baFe-iF' layi' f =wiyh=) a) f ey' 4he' l') ical=da) pe4=af d= fhzzy=c' f y4' l,=*SSSuSSuSaSn onSSoSanu HJaSSA MoniSoSina,*=4(3),=(=f f 7)=384-4f ( )=

y h==),=y=d hh,=d)=c h' ,=f evel' p) ef y=' f=a=Fl' yleFF=yhbhla4=lif ea4=if ye4i' 4=pe4) af ef y=) a) f ey=) ic4' -) ' y' 4f' 44' b' yic=applicayi' f F,=*IEEE oSanSaSSionS on MaanJSS,*=4=(=f),=(=f f ( )=3=88-3==f)=

= acy aychy=P)=S),=b acS) af =P),=b ') af =y),=A=qhaf yiyayive=) a) f eyic=b4aSif) =expe4i) ef y,=*AS JSiSan JouSnaSof SAySiSS,*=6=(==),=(===3)==f =6-==f =)=

= ajh) da4=S)==),=Oil=hyd4ahlic=FyFye) F,=p4if cipleF=af d=) aif yef af ce,=*MS Sau iHiSS S SofJSSionaS,*= =f f =)=

= a4) ' liF=f)=y),=Pylee=y)=y),= 4' vay=f),= eave=) ' de=dyf a) icF=' f=a=y4acSed=ai4=chFhi' f =vehicle=wiyh= Fe) i-acyive=ai4ba) =Fec' f da4y=FhFpef Fi' f ,=*ASME JouSnaS of S ynaS iS SySSJS S, MJaSuSJS JnS, anu n onSSoS,*=7(4),=(==7( )=3===-4f 7)=

= a4yif F=I),=EFyeveF=y),=da=Silva=F)=P),=S e4delh' =P),=Elecy4') a) f eyic=hyb4id=acyive-paFFive=vehicle= FhFpef Fi' f =FyFye) ,=*IEEE VJAiSuSaS oJSAnoSoay n onfJSJnSJ,*=3,=(=====)===73-==77)=

= a4yif F=I),=EFyeveF=y),= a4qheF=c)=f),=da=Silva=F)=P),=Pe4) af ef y-) a) f eyF=lif ea4=acyhay' 4F=applicabiliyy= if =ahy') ' bile=acyive=FhFpef Fi' f F,=*IEEE oSanSaSSionS on VJAiSuSaS oJSAnoSoay,*=(((=),=yaf ha4y= =f f 6)=

=

= e44iyy-P)=f),=PaFichif FSyj==)=y),=y if ea4=4ecip4' cayif) =elecy4ical=) ef e4ay' 4,=*S Sa SaSJnS S uS bJS 1, 88818*=(==8())=

= illiSef=6)=F),=Acyive=ShFpef Fi' f,=*SAE oJSAniSaS SaEJS S oa 888899,=WaSSJnuaSJ, SA: SoSiJSy of AuSoS oSvJ EnainJJSS*,==88)=

= iyFchSe==),=If flhef ce=' f=4' ad=af d=vehicle=di) ef Fi' f F=' f=yhe=a) pliyhde=' f=b' dy=) ' yi' f F=af d= dyf a) ic=wheel=l' adF,=*SAE, SSJESinS*3=f P ,==6=)=

= izhf ' =P),=Iwada4e=),=E af aha4a=),=' ' ya) a='),=Af zai=P),=E i4ei=),=d a) ada=),=P ' f Fide4ayi' f F=' f= elecy4ical=af d=) echaf ical=yi) e=c' f Fyaf yF=' f=a=) ' vif) -) a) f ey-yype=lif ea4=f P=) ' y' 4,=*SJnSoSS anu ASSuaSoSS*,=8=,=(=f f f)=3f =-3f 4)=

= ' ' 4e=y)=y)=Palazz' l' =A)=b)=c adaf) i==)=')=' aFcaS=A)=F)=b 4' wf =c)== ' f ya) he=c),==yb4id=f4icyi' f - i) pacy=vib4ayi' f =da) pe4=f' 4=c4y') ef ic=4' yayif) =) achif e4y,=*JouSnaS of ESoEuSSion anu Eou JS*,= =3(4),=(===7),=4=(=(f =)=

= h4yy=b)=S),=Elecy4ic=va4iable=da) pif) =vehicle=FhFpef Fi' f ,=*S SaSaSJnSS nuS bJS 181, , 8,* =(==8=))=

E a) aya='),=Sh) ih4a==),=A=) eyh' d=f' 4=' byaif if) =a=lif ea4=Fp4if) =f' 4=a=pe4) af ef y=) a) f ey=leviyayi' f= FyFye) =hFif) =elecy4') a) f eyic=c' f y4' l,=*EEE oSanSaSSonS on MaanJSSS*,=3(3)=(===( )=33=-=338)=

E ' 4=')==),=A4' f==),=6 ij' f ' ,=f eFi) f =' f=a=(=S6 =yhbhla4=pe4) af ef y=) a) f ey=lif ea4=) ef e4ay' 4,=*39SA EnSJSnaSonaSS nivJSSiSJS S ou JS EnainJJSina n onfJSJnSJ, S S En 1881*,=(=f f 4)=(=8-(3=)=

Payy=P)=y),=f eFi) f =af d=yeFyif) =' f=a=c' axial=lif ea4=) a) f eyic=Fp4if) =wiyh=if ye) 4al=lif ea4=) ' y' 4,=*EEE oSanSaSSonS on MaanJSSS*,==(( ),=(==8( )=7(=-=76=)=

P' laS=P),=b h4y' f ==)=P),=I) p4' vif) =FhFpef Fi' f =da) pif) ,=y' h4f al=' f=Ahy') ' yive=Ef) if ee4if) ,= Feb)==7=,=3-=7)=

P' lla4d==)=c),=Acyive=FhFpef Fi' f F=ef haf ce=4ide=qhaliyy,=*SaiSu ay aWJSSJ EnSJSnaSonaS*,=3=,=(==83)= 8(f -8(3)=

P' lla4d==)=c),=Si) ' f F=E)=y)=A),=PaFFef) e4e') f' 4y=yhe=4' le=' f=acyive=FhFpef Fi' f F,=*S SoSJJuinaS of SAJ EnSSSuSon of MJSAaniSaSEnainJJSS, SaSSS* ,==8=(==),=(==84),=6=-=7( )=

P4eh) ' f y=A)=S ib4ayi' f =c' f y4' l=' f=acyive=Fy4hcyh4eF,=*S Suu JS ASauJS iS SubSSAJSS S oSuSJSAS*,=e-b ' ' S= ISb E =f -3f 6-484==-6,=(=f f =))=

= =

=

P4' S' p=c),=Sha4p=)=S),=Pe4f' 4) afce=efhafce) ey=' f=Ii) iyed-bafdwiyh=acyive=ahy') ' yive=FhFpefFi' f F= by=4' ad=p4eview,=IEEE SSoSJJuinaSin onSSoSoAJoSy anu AEESSaSonS,=4=,=(==( )=4f-=48)=

=aSheja=S),=Saf Sa4=S),=S ib4ayi' f=afd=Fh' cS=iF' layi' f=pe4f' 4) afce=' f=a=Fe) i-acyive=4' f-' ff'=da) pe4,= JouSnaSof VibSaSon, ASouSSSS, SSSJSS, anu SJSabiSSy in S JSian,=f 7(4),=(==8( )=3=8-4f 3)=

=aSheja=S),=Sh==),=Saf Sa4=P)=S),=AfalyFiF=' f=a=paFFive=Feqhefyial=hyd4ahlic=da) pe4=f' 4=vehicle= FhFpefFi' f,=VJAiSSJ SySSJS S ynaS iSS,==((),=(==f)=8=-3==)=

=hifef4afS='),=A) a) l' h=E)=b),=y' haffe=A),=6 allace=A),=P4hdell=y),=' i) ble='),=AillF=y),=Sh) idy=E),= Phaf=P),=Sweefy=b),=Schache4=A),=E' vel=' ceaf=efe4) y=pe4) afefy=) a) fey=lifea4=) efe4ay' 4=bh' y,= SJnJu abSJ EnJSay,=3==(=ff 6)==7=-==8)=

=yba=f),=I) p4' ve) efy=if=dyf a) ic=cha4acye4iFyicF=' f=ahy') ' bile=FhFpefFi' f=FyFye) F=(Pa4y=I)=Pw' -) aFF=FyFye) F),=VJAiSSJ SySSJS S ynaS iSS,=3(=),=(==74)=7-4( )=

Sch) id==),=S a4) a=P),=AfalyFiF=' f=vib4ayi' f-iF' layif) =FyFye) F=f' 4=Fcaffif) =yhffelif) =) ic4' Fc' peF,= S SSSaS iSSoSSoEy,=4=-44,=Pa4y=b =(===)=6=f-=6=( )=

Se) al=y),=h i' -Pe' =y),=S ehichla4=4eFiFyafce=y' =) ' yi' f=aF=ifflhefced=by=4' ad=4' h) hfeFF,=AuSSSaSan Soau SJSJaSSA,===(4)=(==8=),===-===)=

Sha4p==)=S),=P4' lla=f)=A),==' ad=vehicle=FhFpefFi' f=FyFye) =deFi) f=-=a=4eview,=VJAiSSJ SySSJS S ynaS iSS,=6,=(==87)=67-===)=

Sha4p==)=S),=aFFaf=S)=A,=Phe=4elayive=pe4f' 4) afce=capabilityieF=' f=paFFive,=acyive=afd=Fe) i-acyive=ca4= FhFpefFi' f=FyFye) F,=SSoSJJuinaS of SAJ EnSSSuSon of MJSAaniSaS EnainJJSS, SaSS S ,==ff (3),= (==86)===-==8)=

Sha4p==)=S),=aFFaf=S)=A),=Phe=fhfda) efyalF=' f=paFFive=ahy') ' yive=FhFpefFi' f=FyFye) =deFi) f,=SoSiJSy of EnviSonS JnSaSEnainJJSS n onfJSJnSJ on S ynaS iSS in AuSoS oSvJ EnainJJSina,===84,=f 4-==( )=

Shef)=d ,=S ehicle=FhFpefFi' f=vib4ayi' f=c' fy4' l=wiyh=) a) fey' 4he' l') ical=da) pe4F,=SAS a oAJSiS, S nivJSSiSy of WaSJSSoo,=ff ( )=

S' daf' ==),=b ae=y)=S),=If) af=f)=y),=b elvif=6)='),=P' fcepy=afd=) ' del=' f=Eddy=ch44efy=da) pe4=f' 4= vib4ayi' f=Fhpp4eFFi' f=' f=a=bea) , JouSnaSof Sounu anu VibSaSon=88,=(=ff ( )==77-==6)=

S' daf' ==),=b ae=y)=S),=If) af=f)=y),=b elvif=6)='),=I) p4' ved=P' fcepy=afd== ' del=' f=Eddy=P h44efy= f a) pe4,=oSanSaSSon of SAJ ASME,==8,=yhf e=ff 6,=pp)==4-3f =)=

= =

Shda=d),=E aSadai=S),=E aSaf ' ='),=Syhdy=' f=yhe=Felf=p' we4ed=acyive=vib4ayi' f=c' fy4' l,=*AouSSA EnSJSnaSonaSn onfJSJnSJ on MoSon anu VibSaSon n onSSoS*(===8)===-=6)=

Shda=d),=Shiiba=P),=A=f ew=hyb4id=FhFpef Fi' f=FyFye) =wiyh=acyive=c' fy4' l=af d=ef e4) y=4e) ef e4ayi' f,= *VJAiSSJ SySSJS S ynaS iSS*,=Fhpple) ef y=(,=(===6)=64=-6(4)=

Shda=d),=Shiiba=P),= i' ='),=' awa) ' y' =d),=P ' f d' =P),=d a) a) aya=),=Syhdy=' f=elecy4') a) f eyic=da) pe4= f' 4= ahy') ' bileF= wiyh= f' f lif ea4= da) pif) = f' 4ce= cha4acye4iFyicF,= *VJAiSSJ SySSJS S ynaS iSS*,= Fhpple) ef y4=,=(=f f 4)=637-646)=

PeFhi) a=),=Paf aSa=),= iya) ' y' ='),=E ' h) hchi='),= if aya='),=Effecy=' f=eddy=ch44ef y=da) pe4F=' f= yhe= vib4ayi' f= p4' pe4yieF= if = Fhpe4c' f dhcyif) = leviyayi' f = hFif) =) ely=p4' ceFFed= d baP hO= bhlS= Fhpe4c' f dhcy' 4F,=*S AySiSS n*),=74,=(===7)=7-=3)=

Ph') pF f=A)=c),=Opyi) h) =da) pif) =if =a=4af d') ly=exciyed=f ' f-lif ea4=FhFpef Fi' f,=*S SoSJJuinaS of SAJ EnSSSuSon of MJSAaniSaSEnainJJSS*,=84(=A),=(==6=)=6=-=78)=

P' f' li=A),=f yf a) ic=Pha4acye4iFyicF=' f=eddy=ch44ef y=da) pe4F=af d=c' hple4F,=*JouSnaS of Sounu anu VibSaSon*,=3f =(3-( ),=(=f f 7)=(76-(==)=

S ahid=O),=Expe4i) ef yal=if veFyi) ayi' f=' f=a=va4iable=da) pe4,=*MJSAu oSJS SySSJS EnS*)=f ' ch) ef y=file= f h) be4=f 6f ==(-=,=yaf ha4y=f f 6)=

S ail' f,=y),=Peyiyeaf ,=b),=F4apa4d,=b),=y ebihaf ,=f),=Acyive=iF' layi' f=if =Fpace=y4hFF=Fy4hcyh4eF=f4') = c' f cepyy' =i) ple) ef yayi' f,=*SS aSSMaSJSiaSS anu SSSuSSuSJS*,=8(6),=(====)=78=—7=f)=

6 af) =y),== ' we=f),=yewell=c)=6),=Af alyFiF=af d=deFi) f =' pyi) izayi' f=' f=af =i) p4' ved=axially=) a) f eyized=yhbhla4=pe4) af ef y-) a) f ey=) achif e,=*EEE oSanSaSSonS on EnJSay n onvJSSion*,=v' l)= ==(=),=yhf e=f f 4)=

6 ayaSabe=),=P ' hd' =),=P hiba=O),=izh) i=E),=EbiFawa=),=Fhjiya=P),= eFp' f Fe=c' f y4' l=pe4f' 4) af ce=' f= a=hyb4id=) aFF=da) pe4=applied=y' =a=yall=bhildif) ,=*EaSSA uaJJ JnainJJSina H SSSuSSuSaS uynaS iSS*,= 3f (==),=(=f f =)=6((==676)=

6 eave4=E)=6),= eaFh4e=' f=Fh' cS=abF' 4be4=pe4f' 4) af ce,=*AuSoS oSvJ EnuuSSSy*,=(====)=87f -=7=)=

6 eb=,=hyyp-SSwww)ca4bibleF)c') SFhFpef Fi' f ybibleyp) =)hy) l,=Ap4il,=f f =)=

6 eb=,=hyyp-SSdelphi)c') S) af hfacyh4e4FSahy' S yhe4SSideS) a) f e4ideS=Ap4il,=f f =)=

=

6 ef=d)='),=AFce==),== eyh' d=' f=4af d') =vib4ayi' f =' f=hyFye4eyic=FyFye) ,=*JouSnaS of EnainJJSina MJSAaniSS S iviSion, ASn E*,=f =(=),=(==76)==4=-=63)=

6 ef del=c)==),=SyecSleif =c)=y),=A=4e) ef e4ayive=acyive=FhFpef Fi' f =FyFye) ,=*SAE oJSAniSaSS aEJS SJSiJS*,= ==f 6(==(====)====-=3( )=

6 heele4,==)=A),=F' 4) hlaF=f' 4=yhe=FSif =effecy,=*S SoSJJuinaS of SAJ EnSSSuSJ of S auio EnainJJSS*,=3f (=),= (==4=),=4===-4=4)=

6 hiye=F)==),=Flhid== echaf icF,=*AifSA EuiSon, Wn S MS Sau HiSS*,==f f 3)=

6 iede4icS==)=f),=c ahyhei4=E),=P a) pbell=f)=A),== ' ch' f =P),== a) f eyic=b 4aSif) -=Si) ple=Phe' 4y=af d= Expe4i) ef y,=*AS JSiSan JouSnaSof S AySiSS*,=((=(6),=(==87)=(f f -(f 3)=

d abhya=),== idaSa=),=FhShFhi) a=E),=f flhef ce= f=FhFpef Fi' f =f4icyi' f = f =4idif) =c') f' 4y,=*S SoSJJuinaS of 8SA EAVSS SyS EoSiuS on SAJ S ynaS iSS of VJAiSSJS on S oauS anu on S aiSu ay oSaSJS, n aS bSiuaJ S S*,===8=,==8==-===3)=

d aji) a==),=6 aSiwaSa==),== if e) iFhi=),=Fhjiwa4a=E),=P a) h4a=),=f eFi) f =' f=lif ea4=f P =) ' y' 4=f' 4= hi) h-Fpeed=p' Fiyi' f if) ,=Sef F' 4F=af d=Acyhay' 4F,=8==(=f f f)==8==-==84)=

d ah==)=P),=P hef =P),==Elecy4ic-hyd4ahlic=acyhay' 4=deFi) f =f' 4=a=hyb4id=Fqheeze-fil) =da) pe4-) ' hf yed= 4i) id=4' y' 4=FyFye) =wiyh=acyive=c' f y4' l,=y' h4f al= f=vib4ayi' f =af d=ac' hFyicF,===8(=),=(==f f 6)==76=83)=